URBAN MACHINES:
Public Space in a
Digital Culture

D1547175

Marcella Del Signore
Gernot Riether

Table of Contents

4   **INTRODUCTION**
5   (URBAN) and (MACHINES)
8   AN OPERATIVE FRAMEWORK
9   Appropriative and Adaptive
10 Prototypical
11 Hybrid and Systemic
12 Performative
13 Generative and Catalytic

16 ESSAYS
18 INTERACTING
18 The changing notion of INTERACTION and the PUBLIC realm
22 INTERACTION and the production of the DEVICE
25 INTERACTION and the production of technologically integrated PUBLIC SPACES

32 INTEGRATING
32 The INTEGRATION of INFORMATION and LOCATION
33 URBAN-WARE for the SMART/INTELLIGENT CITY
37 The EXPERIENCE of DIGITAL URBANISM

42 EXPANDING
42 The EXPERIENCE of EXPANDED Urban Space
42 Urban DEVICE and Urban SOFTWARE
50 STRATEGIES to expand URBAN SPACE

54 HACKING
54 HACKING Urban Space
61 OPEN DEVICE and CODE
63 OPEN SOURCE City

70 NETWORKING
70 NETWORK structures and emerging URBAN FORMS
72 The (social) PRODUCTION OF SPACE through the NETWORKED PUBLIC
75 LOCAL - GLOBAL / ACTIONS / AGENCIES

84 CASE STUDIES
86 ECO-BOULEVARD, AIR TREE
    Ecosistema Urbano

94 COSMO
    Andrés Jaque
    Office for Political Innovation

102 DATAGROVE
    Future Cities Lab

110 WENDY
    HWKN

118   iLOUNGE
      Marcella Del Signore (X-Topia) + Mona El Khafif (ScaleShift)

126   LUMEN
      Jenny Sabin Studio

134   LIVING LIGHT
      The Living

142   iWEB
      ONL / Kas Oosterhuis

150   BUBBLES
      FoxLin

158   SKIN
      Gernot Riether + Damien Valero

166   CLOUD
      Single Speed Design

174   BLUR
      Diller Scofidio + Renfro

182   SERPENTINE GALLERY PAVILION
      OMA - Office for Metropolitan Architecture

190   DIGITAL WATER PAVILION
      Carlo Ratti Associati

198   CONVERSATIONS

200   URBAN SOCIAL DESIGN, NETWORK, PARTICIPATION and OPEN SOURCE CITY
      Ecosistema Urbano

208   SENSEable CITY
      Carlo Ratti with Matthew Claudel

214   SYSTEMIC ARCHITECTURE
      ecoLogicStudio

226   CITIES, DATA, PARTICIPATION and OPEN SOURCE ARCHITECTURE
      Usman Haque

236   SITUATED TECHNOLOGIES
      Omar Khan

244   SYNTHETIC URBAN SCENARIOS
      Areti Markopoulou + Manuel Gausa

252   MACHINES and APPARATUSES AS SCENARIO
      François Roche

264   **ACKNOWLEDGEMENTS**
264   **ILLUSTRATION CREDITS**

INTRODUCTION

"Public space is a place where many activities overlap: rich confusion, commerce, seduction, and filth. Public space works not as a designed element, but it is instead carved out by wheeling and dealing, crossroads, and the chance at freedom, where a person emerges from shadows into light that grows into the ever-extending space of public gathering and demonstration and seeps into every open pore of the city."[1] Aaron Betsky

"This is merely to say that the personal and social consequences of any medium—that is of any extension of ourselves—result from the new scale that is introduced into our affairs by each extension of ourselves, or by any new technology."[2] Marshall McLuhan

Over the last few decades an increasingly collaborative work developed among spatial practitioners such as architects, urban planners, artists and media designers has produced a particular landscape of projects that engage information technology as a catalytic tool for expanding, augmenting and altering the public and social interactions in the physical urban space.

As public space is increasingly restricted not only as physical space but also in terms of autonomy and manifestation of the social realm, the interventions that will be discussed in this book reclaim people's common right to public space. For spatial practitioners, a series of opportunities arise from the possibility of engaging networked digital technologies as catalysts for processes that might have a strong impact on social, cultural and environmental future scenarios.

The line of inquiry—set up through the research framework of the book— starts by reading the contemporary urban space as a set of shared, common, sentient and networked conditions. The book will show how information technology and digital media are used as tools for the making of urban places. In the first part of the book, interacting, integrating, expanding, hacking and networking are defined as categories to explore modes of operations in public space and the urban environment. In the second part, projects and prototypes are presented to dissect different modes in which spatial practitioners operate. The third part of the book features interviews with architects, urban planners, artists and media designers to provide a series of key themes to expand on the impact of networked digital technologies on future urban scenarios.

The notion of the machine—as suggested in the title of the book—is derived from Gilles Deleuze and Félix Guattari and not intended as a mechanical construct, but on the contrary, envisioned as the potential of machines as assemblage of different devices that generate a "functioning whole" and "connected system." The assemblage in this scenario is intrinsically related to agencies that have potential for actions played out in the physical urban space.

Design practice is increasingly supporting the relevance of integrated models that take into account not only space but also systems. Through the essays in the first part of the book, case studies in the second part and interviews in the third part, the book explores the proliferation of those systems in urban spaces providing a glance to scenarios that might emerge in the near future. The topics presented suggest a vehicle for discussion and moment of reflection on the impact of those themes on contemporary spatial practice. The cross disciplinary nature of the case studies, along with the categories of investigation engaged throughout the chapters, make this book an important resource for architects, urban planners and digital media artists as well as academics in these disciplines.

(URBAN) and (MACHINES)

While during modernism the city was structured on replication of coherent forms and divided into functional pockets, contemporary urban conditions are more complex. The contemporary city is positioned between complexity and immediacy.[3] As information technology becomes pervasive, one has to rethink the rules for communication between the citizen and physical urban space. Emergent technologies have brought into question the role of the material city in representing the public and the collective experience of urban space.[4] Information and matter, code and space collapse into a new system, and mediated spaces become an architectural problem.[5] In this new scenario, the role of machines is intended as a set of devices that become relevant to the experience of urban space and the public realm.

A device can be defined as an apparatus, instrument or tool designed for a specific purpose. Devices perform, inform and continually transform the environment and our perception. They can manipulate data by seeking performative spatial relationships. Accordingly a machine can be defined for its inherent meaning of being a system of devices able to communicate and perform within a certain environment. A machine implies the notion of "something that has been constructed" and "function with a specific purpose" while being composed by parts that respond to a "functioning whole." The notion of assemblage developed by Deleuze and Guattari is intended as a composition of heterogeneous elements that give rise to a new system. In Deleuze's book *A Thousand Plateaus* the concept of machine can be understood as a more complex formulation of the concept of assemblage. In Guattari's terms, a machine is a composition of heterogeneous elements—subjective, social, technical, spatial, physical and process-related—that delimits a series of conditions for the production of the real.[6] This entails that the notion of machine is directly connected to the concept of agency as it has the capacity or potential for an action within an environment. Rather than seeing those interventions only as objects or installations, they are seen as assemblages whose spatial parameters merge with information, networks, devices, media and users to create a public space that is more responsive, participatory and collective. The inhabitants are a dynamic part of the assemblage and become active producers of the public space. Space in this scenario emerges as a social product,[7] and citizens are empowered in the production of it. Those interventions can be seen as tools to produce hybrid public spaces that in the temporary, interim or permanent phase are active, networked and responsive. Urban Machines in that way provide an alternative scenario to the production of physical public space.

From this conceptual context, Urban Machines are interventions in the physical urban public space that function as a system or set of devices, and through information technology, mediate the relationship between the urban environment and the user. In this framework the spatial practitioner is designing and programming these machines to promote, test and prototype the relation between city, technology and the human scale. The machines have the potential to generate a new type of either permanent or temporal urban geography operating as largescale plug-in systems. CJ Lim, in his book *Devices,* argues that a technological or abstract understanding of such machines and their construction can influence and redefine the potential for architecture and spatial thinking.[8] Urban Machines are a family of projects designed and developed to mutually enrich relationships between people, the space they inhabit and the urban environment. As the city is more and more produced through entropic processes, Urban Machines could operate as synthetic systems determined by the recombination of multiple parameters into one performative spatial form (fig.1). Those systems perform differently based on how and where they are situated, their scale and their relationship to the environment. Urban Machines could help to intensify the interaction between urban space, people, objects, architecture and media devices. They have the potential to test and prototype models for future urban scenarios.

Fig.1: Active Public Space – Glories Regenerative Systems; IaaC, 2015/2016.

The concept of spatial machines that operate at the scale of the city while creating a direct exchange with the public is not new. One can trace throughout history examples of visionary urban machines that pushed the notion of the urban imagery (fig.2). In 1964 Cedric Price conceived the Fun Palace as a hybrid system that resided somewhere between a participatory space and an unfinished infrastructure. As a provocation, the Fun Palace was envisioned as a large performative machine where architecture enabled the users to reconfigure modes of occupation. Provisional space, the endless process of construction, the dismantling and reassembly comprised the machine-like system. The inhabitants had to adapt and change because of its incompleteness. The Fun Palace was conceived more as an instrument rather than a building. At the same time, Archigram's Walking City was designed as a complex macro-system that had the vision to absorb the city and transform it into an adaptive and dynamic living being. Devices with a specific purpose were synthesized into a "walking creature" to provide an alternative model of living. Using similar parameters, Toyo Ito's Tower of Winds was a machine that became an urban icon and performed as a public signal by responding to the changing conditions of a nearby underground train station.

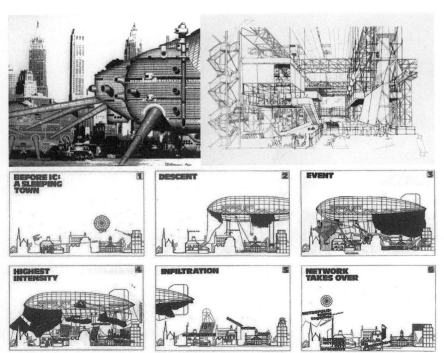

Fig.2: Archigram, Walking City, 1964; Cedric Price, The Fun Palace, 1964; Archigram, Instant City, 1964.

AN OPERATIVE FRAMEWORK

The operative framework of Urban Machines is defined by common categories shared by the types of interventions discussed in the book. It identifies possible overarching principles to position their mode of operation in the urban context. The projects presented have been selected for their shared capacity of being systems, that through the input of data, information technology driven processes and the production of outputs, can impact how modes of experiencing the public realm are imagined. Some are more temporary or interim than others, but all of them share the potential of providing a model that could be tested and possibly implemented in the long–term in our cities.

The shared categories are:

1. APPROPRIATIVE and ADAPTIVE
2. PROTOTYPICAL
3. HYBRID and SYSTEMIC
4. PERFORMATIVE
5. GENERATIVE and CATALYTIC

1. Appropriative and Adaptive

Urban Machines are temporary or permanent spatial acupunctures or plug-ins in the urban environment. Urbanist Manuel de Solà-Morales describes urban acupunctures as catalytic strategies for urban renewal where small interventions, realizable in a short period of time, are capable of achieving a maximum impact in the immediate urban space (fig.3). Because of the short timeframe in which they are developed, those types of interventions are similar to injections, able to heal rapidly the immediate area and bring new life.[9] They appropriate the city, subverting the traditional order of urban processes developed by top-down approaches, changing and questioning the culture of planning by providing an alternative form of urban occupation.

Fig.3: Jaime Lerner, Urban Acupuncture, 2007; Atelier d'Architecture Autogérée, Tactics for Resilient Post-Urban Development, 2014.

In the late 1950s the Situationists moved away from functional urban planning and started looking at the social realm as a field for urban production. Their psycho-geographical explorations began by understanding people in space and how they were actors in producing the social scale of the city. The city was not perceived as a static, planned and controlled object but a space that could be written and rewritten, following a cyclic process of construction and decay. Time and space follow a continuous process of re-creation because of the city's social dimension. While the notion of the temporary city is influenced by the Situationists, temporality relates to more recent approaches in urban planning. Margaret Crawford in her book *Everyday Urbanism* says: "It is an approach to urbanism that finds its meaning in everyday life, but in an everyday life that always turns out to be far more than just the ordinary and banal routines that we all experience." In contrast to urban planning, *Everyday Urbanism* looks at people, situations and places resulting at a high level of specificity in how these spaces are constructed.[10]

As temporary urbanism provides an alternative form of urban planning, it proceeds tactically rather than strategically. As designers are immersed in a phase where formal master plans are difficult to develop and to implement due to the lack of resources, more flexible, short-term initiatives are undertaken to accelerate urban renovation. More and more, instant, pop-up, interim, flexible, temporary interventions are seen as a rapid bottom-up response to what the city needs at the moment, to resolve the contingent urban issues, to produce urbanity without long-term implementations. Those projects accept the inherent transient condition of urban life. Urban Machines can be powerful tools to foster bottom-up and open source urban processes promoting urban renewal.

2. Prototypical

The notion of prototype is directly connected to the notion of testing. Prototypes entail a working model, an early sample built to test a process; they are instances for learning and optimization and are process-oriented. Testing establishes a series of procedures for critical evaluation and development. Both prototyping and testing are intrinsically connected and negotiate between "process" and "product" (fig.4).

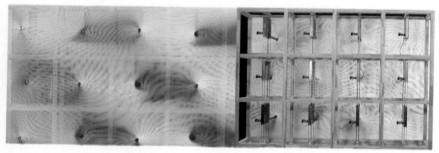

Fig.4: Barkow Leibinger, Kinetic Wall, prototype exhibited at the Venice Biennale 2014.

Urban Machines function as prototypes and test a system of relationships. Functioning as prototypes they seek to engage the public or a specific urban condition. Once a strategy has been proven successful the prototype might inform long-term implementations. This has a strong potential to replicate the same intervention in other urban contexts with the possibility of being adopted in the city on a long-term basis. The process of prototyping, replicating and adopting can promote urban innovation. Urban prototyping as a movement is exploring those processes demonstrating that participatory design, art, and technology can improve cities.[11]

3. Hybrid and Systemic

Hybrid public space refers to collectively inhabited urban space that is traversed by digital flows of data and images that enhance and alter the traditional interaction between the body and its physical, social and symbolic environment.[12] Those interventions generate a hybrid space that acts as a catalyst for urban activities. In the mix of people, flows, networks, data and electronics, new relationships between humans and machines, time and space can be found. An urban space that embraces technologies changes everyday rituals and how social interactions are mediated. At the same time, the importance of the physical encounter must be recognized. Those interventions embody the capacity of bridging the physical and non-physical into a hybrid condition (fig.5). They merge the hardware (space, tectonics, materials) and software (information, systems, networks).

Fig.5: Augmented Structures v1.1: Acoustic Formations by Alper Derinboğaz, New Media art by Refik Anadol, Istanbul, 2011.

Urban Machines produce new ecologies in the way Guattari theorizes: environmental or technical ecology, social ecology and mental ecology. Urban Machines are an urban heterotopia, a public space that emerges between environmental conditions, ambient conditions and situations, an urban manifestation of event and memory, a space that is temporary but leaves a permanent transformation in its urban context. Experimenting at the intersection of information technology, urban space and architecture, Urban Machines emphasize hybridity over mono-functionality. Environment, space, technology and different forms of use are intertwined to produce a space that encourages new modes of urbanity and the emergence of new forms of public life. As a space-environment Urban Machines allow for multiple conditions to exist simultaneously. Urban Machines promote a continuous hybridization and exchange between the city and its citizens, place and technology.

4. Performative

As the nature of the public environment has changed dramatically, Urban Machines seek to create atmospheres that embody this negotiated status, engaging the public in constructed agencies.[13] The same perspective was theorized by Henri Lefebvre who pointed out that the city is both a product and a medium created by social praxis and socio-spatial processes.[14] Lefebvre decodes the urban space in three dimensions: 1) the perceived space: space produced by the collective activities in the urban space; 2) the imagined space: space constructed by urban planners and architects as a "representational space" projected onto the reality; 3) the experienced and suffered space: space experienced by users and mediated through images and symbols of everyday life.

Urban Machines are systems able to extend material space into space for action. Action is a generating mechanism to express form and space. Those interventions have an inherent relational nature and the ability to set up a public system for interactions and events to occur (fig.6). Urban Machines construct scenarios in which the public is invited to enter a manifold space where the experience is multilayered and set in motion by a series of spatial, ephemeral and technological mediated devices. They intervene in the public realm as systems that are socially, technologically and physically integrated.

Fig.6: Rafael Lozano-Hemmer, Under Scan, Relational Architecture; Lincoln, 2005 and Leicester, 2006.

5. Generative and Catalytic

One interesting aspect of Urban Machines is the way city planning can learn from them or the way they can become a model to catalyze processes while generating scenarios for long-term implementation. By understanding the main actors that operate in the city, from the municipality to economic forces, those interventions can embody powerful catalytic urban tactics. They are initiators of temporary uses that can provide a dialogue between the site where they are inserted and the local actors involved (fig.7). On some level, the failure of some traditional top-down master planning approaches opens up the potential of interim insertions for the capacity of accelerating urban renewal in a short time. The fact has to be acknowledged that a city is a living organism that is constantly changing and that finalized or rigid conditions are never attainable or desirable. Such catalytic tactics enact a series of actions by users or residents through empowerment. The potential of temporary plug-in spaces is to act as a generator that attracts forces, from the social to the economical level. Such spaces begin to implement programs at the micro level. The same programs implemented through some traditional planning tools could take a longer time or never be implemented.

Fig.7: People's Architecture Office, People's Canopy, Preston - Lancashire, UK, 2015.

Urban Machines can be the accelerators of multiple urban operations. They are dynamic agents able to influence urban configurations and narratives yet to come. The generative potential of these interventions is the capacity to catalyze processes of creation of the "open city," a city that is in constant evolution and that can be transformed through bottom-up and overlapping of functions, while initiating processes that start the dialogue about urban visions in the surrounding community.

The above operative framework provides shared conditions for the deployment of those forms of urban interventions in contemporary cities. As deep processes of transformation are changing cities, spatial practitioners will have to expand the tools at their disposal to address increasing urban challenges. Through essays, case studies and interviews this book presents a snapshot of the current practices and modes of operation in the urban environment. The debate and discussion are open, and the book is presented as a tool of inquiry on present and possible future scenarios.

1. Betsky, A. (1998) 'Nothing but Flowers: Against Public Space', in Bell, M. and Leong, S.T., *Slow Space*. New York: Monacelli Press, pp.456-78.

2. McLuhan, M. (1964) *Understanding Media: The Extensions of Man*. Cambridge, Massachusetts: MIT Press.

3. Palumbo, M.L. (2000) *Electronic Bodies and Architectural Disorders*. Berlin: Birkhauser, pp.31.

4. Boyer, C. (1996) *Cyber Cities*. Princeton: Princeton Architectural Press.

5. Manovich, L. (2002) *The Language of New Media*. Cambridge, Massachusetts: MIT Press.

6. Deleuze, J. Guattari, F. (2000) *A Thousand Plateau*. Minneapolis: University of Minnesota Press.

7. Lefebvre, H. (1992) *The Production of Space*. Oxford: Wiley-Blackwell.

8. Lim, CJ (2006) *Devices*. Oxford: Architectural Press.

9. De Sola Morales, M. (2008) *A Matter of Things*. Rotterdam: NAi Publishers.

10. Crawford, M. (2008) *Everyday Urbanism*. New York: The Monacelli Press.

11. Urban Prototyping Research Lab (2012). Available at: http://urbanprototyping.org (Accessed: 15 June 2018).

12. Moreno, S., De Lama, J.P., Andrade, L. H., *Wikiplaza* (2011). Available at: http://ia600707.us.archive.org/9/items/WikiplazaRequestForComments/WikiplazaRequestForComments.pdf (Accessed: 15 June 2018).

13. Bohme, G. (1998) *Atmospheres as an Aesthetic Concept*. Daidalos, Vol. 68, pp.112-115.

14. Lefebvre, H. (1992) *The Production of Space*. Oxford: Wiley-Blackwell.

ESSAYS

1. INTERACTING

2. INTEGRATING

3. EXPANDING

4. HACKING

5. NETWORKING

INTERACTING

THE CHANGING NOTION OF INTERACTION AND THE PUBLIC REALM

The term "interaction" defines an action that emerges from two or more entities that affect each other. Interaction suggests a conceptual framework for the relationship between citizens and the dynamic complexity of public spaces. Gordon Pask[1] theorized in his book *Conversation Theory* that interaction can be generated from a feedback loop that links input, process, and output. "Urban public space" is by definition a spatial manifestation of interaction. It is defined as space that is open and accessible to all members of the public in a society[2] and a space that emerges from the negotiation and dialogue of social, economic, and political forces. As such, it is the city's potential to provide a common ground for public expression and actions to occur. In the context of a public space that is increasingly defined by information technology, its concern must go beyond this spatial and physical definition. How can public space be the domain where both physically and digitally mediated forms of interactions can be revelled and manifested in an increasingly networked society?

Before radio, TV, and Internet existed, people went to public spaces when they wanted to access information about their community to debate political matters, or to engage in commercial trade (e.g. acquiring goods required entering or crossing a public space). Nowadays the Internet provides a constant stream of information and serves as a platform for building communities, is used in politics, is a tool for trade, and facilitates the acquisition of almost anything online.

Before radio, TV, and the Internet existed, villagers gathered in public spaces when they wanted information or had to obtain goods. They gossiped and engaged in heated public debates. They bartered and traded goods and developed relationships in these spaces. Today, desired information is delivered 24/7 via a diet of individually tailored Internet services such as Hulu, Pandora, and a host of others. People debate each other anonymously in the comments sections of local newspapers. Friends on Angie's List tell each other what goods they need, and Amazon will deliver them to their door. Step by step, public space has ceded the exclusiveness of its essential role, a development Walter Benjamin did not foresee when he wrote in the early 19th century that technologies and mass media would extend our public space.

Mass media in the early 19th century started to reactivate and reshape public spaces. When the newspaper was a luxury item, it activated public space in unexpected ways. People took their newspaper to the local café, which became a place for the exchange of information, public discussions, and community building.[3] The cafés linked new media to public space, changing the character of the urban environment. New media was expected to maintain the reader's impatience, which demanded new excitement every day. In his short essay "The Newspaper,"[4] Benjamin illustrates how public space changed into a space characterized by citizens shopping for facts. He noted that while the standard novel was written for the individual reader, the newspaper was a new kind of media written for the masses of cities. Private space became the "space of the story" and public space the "space of facts and advertisement."

When newspapers introduced the letter to the editor, each reader could turn into a writer.[4] The medium itself became a space of interaction, an interaction that required participation, commitment, and citizenship. In 1895 the first movie screening took place in a café in Paris. The new medium of film was used to introduce the novel to the public space. As the letter to the editor turned the newspaper reader into a writer, the medium of film put the audience in the position of the camera,[5] blurring the boundaries of observer and observed, authors and readers, producers and spectators. Public space developed from a space of reality into a space of reality and illusion. Walter Benjamin imagined that the newspaper, film, and other new technologies would continue this trend and change public space into a space of heightened interaction.

When consumption became linked to urban space, public space became a space of the spectacle. With the rise of consumer culture and the entertainment industry, storefronts and arcades expanded. Entering a public space meant "to participate at" and "to be incredibly linked to" the constantly changing spectacle and entertainment of urban spaces. Dress well because you never know what will happen. Walter Benjamin coined the term Flâneur to describe this new behavior of using public space. Flâneur, he said, changed Paris from a generic urban space to a "landscape made up of living people."[6] Franz Hessel described it as a space of interaction in his book Spazieren in Berlin (Walking in Berlin).[7]

Modernism compartmentalized urban space, negating Benjamin's vision of a "much more fully" integrated city."[8] Large shopping malls kicked smaller shops and stores out of business. Radio and TV disconnected the individual from public spaces. Movies were made in studios and displayed in the controlled environments of movie theatres. Music was recorded in studios for hi-fi listening in private spaces.[9] If pre-modernism was about interaction, modernism was about segmentation. In the essay "The Mediated Sensorium," Caroline A. Jones argues that the increasing technology of media contributed to the segmentation of perception.[10] The media philosopher Vilem Flusser states that the development of spaces requiring separation is closely linked to the genesis of technology.[11]

Since artists, designers, and technologists started to experiment with digital information technology in the 1960s, many architectural, urban, and art interventions have promoted interaction as a concept for architecture and the city as space-environments. These interventions tried to reverse the trend of separation by testing new models of interaction. Cybernetitian Gordon Pask and architectural groups such as Archigram suggested more participatory urban spaces where the participants played an important role in the creation of urban space itself. These ideas promoted collaboration and involvement rather than distanced and detached observation. The concept of interaction was affirmed by early computation and seen as a strategy to reconnect what was separated and segregated through modernist architecture and planning. In the 1963 Living City exhibition in London, Archigram explained the city as a "sum of its atmospheres."[12] Archigram used the idea of interaction as an alternative model to urban space that was defined by demarcation and boundaries. Archigram's Instant City was a transient event in which the formal articulation of the space was irrelevant. Its dynamics, relations with people, and the functioning of the environment comprised by many parts and sub-parts were the spatial conditions worth aiming for. The Italian design group Archizoom talked about the city as a flexible environment, a place of activity, and in constant construction. For Archizoom the links between the city's parts were more important than the parts themselves. The city was not understood as a sum of parts but as a system of interactions between parts. Archigram described this as a "living city." Archizoom called it a "no-stop-city."[13]

Pask's Cybernetic Serendipity in 1968 was one of many exhibitions that explored computation to generate environments as new forms of interaction. Sharing Benjamin's vision, Pask speculated that urban space would transform through new media into a space of eventful play "intensified interaction." The trigger of this development would be digital and information technology. Pask understood the generation of forms as an outcome of a digitized interaction between participants and their environment. He defines interaction as a productive process. If the relation between public space and occupant is defined by interaction, the occupant of public space is not just using public space but is at the same time constructing it.

In the 1960s the idea of interaction as a productive mechanism was present across disciplines. The French New Wave movement in film is one example. Rather than producing a film in a closed, controlled movie theater, directors such as Jean-Luc Godard insisted on shooting their films in public spaces. *Breathless*, for example, was shot with a hand-held camera that enabled the director to shoot in the city of Paris without blocking off streets or asking for permission. Curious passers-by who looked into the camera were not seen as problematic but as essential to producing a quality of spontaneity lacking in isolated environments.

The counter movement of the 1960s did not last long. Movie productions moved back into studios and theaters, and stores moved into shopping malls. Archigram's visions of interactive cities and Pask's experiments to use computation to inform interactive urban environments were not realized. The interface of the new digital technology was developed for the private space instead and named "personal computer." Instead of Archigram's visions of cities that emerged from social interactions, individual star architects and the celebration of the spectacle of individual buildings started to dominate the architectural landscape while urban space turned into a backdrop. The personal computer changed computers from specialized to universal machines. A personal universal interface moved people from public to private spaces. Interfaces were made for the individuals and didn't connect to physical public spaces as suggested by Pask. But one could make the argument that many of Pask's concepts of interaction and public space were realized within the Internet.

Digital media reinvented, resized, and restructured communities that flourished exclusively in the digital space, its globalization holding more potential for societal change than television and radio ever could. Since the last century, digital communication technology has changed the social space, the market space, the space of education, and the space of politics. Digital space with all its social networks, blogs, and online forums has become a substitute for interactions in physical space. As described in Levy's analysis of cyber-democracy and information capitalism, the Internet created the bodily collective but lost its connection to physical urban space.[14] In her book *Alone Together: Why We Expect More from Technology and Less from Each Other,* Sherry Turkle describes how technology has become the space of facts and illusion.[15] In the book *Bowling Alone: The Collapse and Revival of American Community,* Robert D. Putnam shows how changes in work and family structures and a suburban lifestyle have contributed to this development.[16]

Pask's innovative idea of a more spatially integrated digital interface, suggestive of a much more intense relation between the digital and physical public realms, had been almost forgotten until Henrik Sjödin introduced publicly accessible wireless LANs at the NetWorld + Interop conference in 1993. What Sjödin proposed became popular during the following years as hotspots. Wireless broadband technology was standardized as Wi-Fi in 1999 and incorporated in most laptops and other mobile devices soon after. Wi-Fi access became available in a growing number of coffee shops, airports, and hotels. With laptops that became more powerful and smaller, people could again share working space with others, a quality that was missed during the era of the desktop. The Internet, like printed newspapers in their heyday, offers access in public spaces and has brought about a renaissance of the café since Starbucks introduced Wi-Fi in all its locations in 2008. As mass media in the early 19th century reshaped public spaces, wireless technology now reshapes those spaces again.

Cafés filled with people working on their laptops present just a glimpse of what is possible with wireless technology. Concepts of interaction embodied in digital technology can now be applied to public space. Wiki, for instance, might soon be picked up as a strategy for urban space to get citizens to interact, participate, and change a public place into a Wiki-platform with a physical dimension. Information technology might change urban space into a space that can be collaboratively changed and constructed by its occupants.

Connecting the physical public space and the virtual digital space suggests that the physical space will turn into a new kind of interface. An urban space that is an interactive interface can become an environment that is highly participatory in nature and is produced by active citizens. As a consequence people in urban environments are not just observers but an integral part of public space. And instead of separating functions as during modernism, urban space can allow for functions to interact. The digital device and the digital city may be seen as a new chance to realize the urban space of heightened interaction suggested by Benjamin, Pask, Archigram, and others.

INTERACTION AND THE PRODUCTION OF THE DEVICE

May 2013: Only 10 percent of computer users accessed the Internet from mobile devices. Seven months later: 67 percent had switched from desktop to mobile devices.[17] In recent years mobile devices have changed from singular types such as cell phones, music players, radios, TVs, video players or laptops to integral hybrid devices such as smart phones. Such devices enable us to move seamlessly from private to publicly shared information and to link social networks, business networks, and other information networks on a single interface. In contrast to the segregation of function in modernism as discussed earlier, today's devices are about the integration and merging of functions. But how can these new devices inform the way people interact with urban space, and how does that affect the behavior of people in public space?

Being permanently connected has already changed the way of interaction: Instead of meeting at a specific location at a specific time, mobile devices allow for people to meet spontaneously in physical space or "meet" digitally. Paper maps are rarely sold anymore[18] and urban Interaction Design[19] has become an emerging field focusing on technologically enhanced urban experiences. These creations blend physical location with digital experience to shape new uses of our cities.

Keith Hampton is one of many exploring how this blend might affect the use of public space. Similar to sociologist William H. Whyte's "Street Life" project in the 1970s, Hampton has used a camera to film public space. As Whyte filmed public spaces over a significant stretch of time to find out where people sit, stand, or converse, Hampton tracks people's use of mobile devices in public space to discover, for example, that cell phone users are five times more likely to linger in a public space. Publishing his results in The *New York Times* in January 2014, Hampton concluded that technology is not driving the society apart anymore.[20]

Understanding interaction as a productive process, a mobile device that is getting progressively smaller, powerful, and integral will also inform new social and cultural practices. In the essay "Interaction Anxieties"[21] Omar Kahn describes the challenges of moving from a desktop interface to new alternative forms of interaction. After the desktop, new models of human-computer interaction were developed. Known as ubiquitous computing, pervasive computing, ambient intelligence, or physical computing, these developments suggest that man-made artifacts and even natural systems consist of hardware and software, and everything can be combined with wireless networking technologies to create an environment in which everything could potentially interact with everything else at any time.

With buildings, places, cloth, and objects becoming computationally augmented the device will become instrumental in constructing productive and provocative relations between people and places, leading to unpredictable results. Projects and prototypes that are currently explored at a small scale, such as an interactive dress designed by J. Meejin Yoon might soon go into production. The Defensible Dress (fig.1) is a garment that protects personal space by using sensor technology to deploy a space-defining physical projection around the body. Operating between public and private realms the dress can be customized to define territories of private space within the public realm. With the increasing possibility of embedding computational intelligence, the concept of interactivity can be used to establish new forms of interaction between people and inform new socio-cultural practices (fig.2). But who will develop the protocols?

In an article on Gordon Pask published in *Architectural Design (AD)* in 2007,[22] Usman Haque critiqued the current pervasive computing approach as one founded on fixed interaction loops, meaning that the way someone is interacting with the tool has been fixed by the designer and assumes that everyone is using it the same way. An interactive pervasive urban space, on the contrary, is not created by extending urban spaces with information technology for someone to "use" but for someone to "participate" in. The device, based on Haque's critique, has to become flexible in allowing people to participate in the construction of protocols themselves. Only in that way will the device allow us to truly interact with urban spaces.

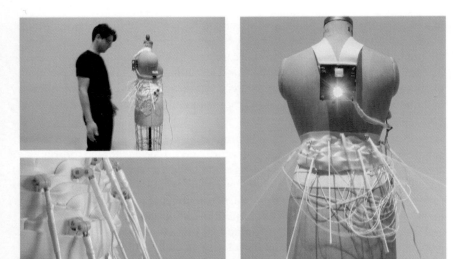

Fig.1: Höweler + Yoon Architecture, The Defensible Dress, 2001.

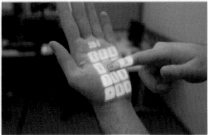

Fig.2: Chris Harrison, Carnegie Mellon University, Skinput, 2010.

INTERACTION AND THE PRODUCTION OF TECHNOLOGICALLY INTEGRATED PUBLIC SPACES

In his book *Placing Words, Symbols, Space, and the City* [23] William Mitchell asks architects and planners to respond to the disconnect between the virtual and the physical, or what he calls cyberspace. Smart phones with digital maps and apps that connect to places in the city have already changed the way of occupying urban spaces. On the other hand, the physical space is slowly adapting to new forms of behaviors. Interaction at the urban scale has a broader sense beyond the pure understanding of the functioning of the device: Urban space will have to be conceptualized as the framework for the public to experiment with new forms of interaction.

The idea of interaction has not only informed digital but also physical space. Computing has progressively moved from the typical desktop interface to new physical and social contexts. Programming environments support an extension of new routines or external objects that are dealing with recognition, analysis, and the generation of new events. Using algorithms to process a large spectrum of possible inputs enables new forms of interaction between a digital and a physical space. New modes of interactive environments have been already tested for interventions at many scales.

Spatial practitioners have tested information technology and the idea of interaction to challenge the relationship between the user and the machine, the viewer and the artifact, or the visitor and the environment. In the context of urban place, redefining the relation between occupant and space will also affect the understanding of what people consider public and what they consider as private space. Allowing an occupant to interact with a space can result in public space that can be turned into a private space. Public spaces will need to be conceptualized as more flexible and dynamic environments that are able to change from public to private or to be public and private at the same time. In such a scenario, the terms public and private will either not exist as categories or will have to be redefined. The boundary between public and private and the static conceptualization of the city that has already been challenged in the 1960s is now challenged again.

Many interactive installations have used information technology to create physical space that is interactive, dynamic, and flexible. A few of them have left the gallery space and moved on to urban settings to redefine user-machine, viewer-artifact or visitor-environment relations. The design of such much larger and complex interactive urban environments requires interdisciplinary collaboration between architecture, engineering, urban planning, and digital media but also the engagement of different groups of interests such as property owners and government. The concept of interaction that has been tested first in small-scale interventions has grown into a new cross-disciplinary challenge to redefine, rebuild, and retrofit public space to function as a framework for new forms of urban interaction.

The practice of Diller Scofidio + Renfro is a good example. The firm first tested ideas as installations and small-scale interventions that later informed architectural and urban scale projects. Their intention to reorganize public space by identifying new types of social interactions and intensifying the correlation between digital and physical space as well as the artificial and natural is clearly reflected in their body of work. The Brasserie (fig.3), their restaurant in

Mies van der Rohe's Seagram Building, uses surfaces and information technology to reconnect what was separated in one of the most distinguished modernist buildings. Besides resurfacing the interior with a wooden skin that integrates the floor, ceiling, and walls, there is also a desire to incorporate digital media to increase interactivity. A live camera connected to a sensor is located at the bar's entrance, taking pictures of arriving guests and projecting them delayed on monitors above the bar. The user does not interact directly with the technology itself, but the technology creates an environment that encourages new forms of interaction through surveillance.

A more recent project from Diller Scofidio + Renfro is the Cultural Shed (fig.4). Close to the Chelsea gallery district in Manhattan, the Cultural Shed manages to define a public space in a 26-acre, column-free mega structure dedicated to the visual and performing arts. Cranes—usually used for loading and unloading large container ships—form a five-story performance space that can expand and contract to accommodate different types of events. The project literally looks like the large hidden space around a theater stage, commonly referred to as fly-space that enables a quick change of stage sets. As such, it

Fig.3: Diller Scofidio + Renfro, The Brasserie Restaurant, New York, 2000.

celebrates the rigging systems, machinery, and infrastructure of the theater stage by removing them from their traditional theatre shell and locating it in the midst of an urban space to interact with the public.

Similarly, two pavilions by OMA-Rem Koolhaas—Prada Transformer in Seoul and Serpentine Pavilion in London—relate to the same category of projects. The Prada Transformer (fig.5), realized in 2009, was designed to transform from a fashion runway to a movie theater to an art gallery and to an event space. The very different spatial requirements informed the planes of a tetrahedron. Depending on the program, the structure, commissioned by Prada for a fashion show, can be rotated by construction cranes to transform itself into a new

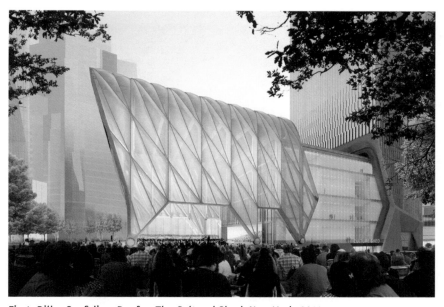

Fig.4: Diller Scofidio + Renfro, The Cultural Shed, New York, 2011.

space. Koolhaas' Serpentine Gallery Pavilion, realized in London in 2006, incorporates a helium and air filled inflatable roof that can be raised and lowered to accommodate different activities and respond to different weather. These projects celebrate an "urban spectacle" determined by the specific interactions that emerge among program, user, space, and context.

In 2011, the city of Singapore completed the construction of Gardens by the Bay (fig.6), a forest of 50-meter high artificial trees. Developed by the British landscape design firm Grant Associates, the "trees" are colossal vertical gardens; their branches are solar panels and rainwater catchers, and their trunks collect water to be reintroduced into the surroundings. The enormous structures form an exotic vertical park, an oasis to escape the everyday city. The

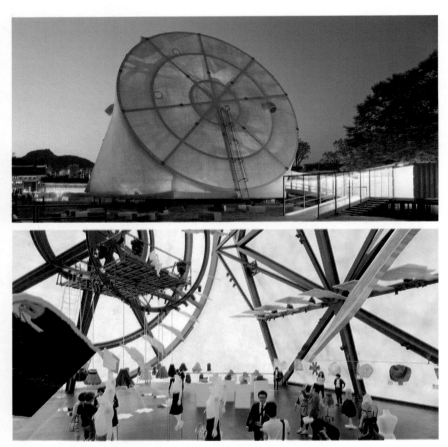

Fig.5: OMA-Rem Koolhaas, Prada Transformer, Seoul, South Korea, 2009.

artificial trees form spaces for festivals and anchor an elevated walkway which provides an alternative way to experience the city. Gardens by the Bay is an urban high-tech intervention that produces energy, collects water, and cleans the air but at the same time defines a large public space.

These projects present scenarios of integration between the users, the environment, and the urban space to rethink forms of interaction for creating public spaces by large scale-machines and interventions. Such projects cannot be defined within conventional categories, as they are permanent but also promote temporary parameters. They are public and private at the same time. They suggest a new quality of urban space that is too dynamic to be categorized as buildings, streets or squares.

Interaction will not only be a concept for urban spaces but also a parameter for designing such places that emerge from collaborative practices. The single building, the shopping mall, the entertainment center or the large private development will pale next to the excitement of these new spaces. Information technology is redefining the dynamic relationships between buildings, urban space, different functions such as stores, places for work, recreation, entertainment and the interplay between public and private.

The production of urban space can be understood as a collective experience in a broader context. Already in 1974 Henri Lefebvre suggested in his book *The Production of Space* that social space is a dynamic space in which interaction is producing the urban space.[24] Designing such a space will be less about problem solving and more about the imagination of shared scenarios. Urban space should be understood as a product of a transformative process that is designed collaboratively as self-organized systems open to new possibilities generated by the users. Information technology is instrumental in the process of transforming an urban space that was characterized by modern segregation for too long into a space of heightened interaction.

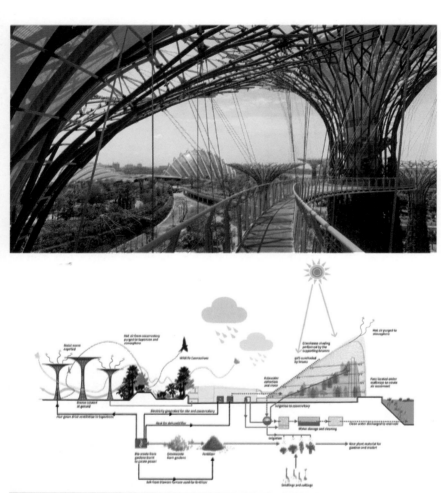

Fig.6: Grant Associates, Gardens by the Bay, Singapore, 2011.

1. Pask, G. (1975) *Conversation, Cognition and Learning*. New York: Elsevier.

2. Neal, Z. and Orum, A. (2010) 'Relocating public space', in Neal, Z. and Orum, A., *Common Ground Readings and Reflections on Public Space*. New York: Routledge, pp. 201–207.

3. Eiland, H. (2003) *Walter Benjamin: Selected Writings*. Volume 4: 1938-1940. Cambridge, Massachusetts: Belknap Press, pp. 13.

4. Eiland, H. and Smith, G. (1999) *Walter Benjamin: Selected Writings*. Volume 2: 1927-1934. Cambridge, Massachusetts: Belknap Press, pp. 265.

5. Arendt, H. (1977) *Illuminations: Essays and Reflections*. New York: Schocken Books, pp. 228.

6. Eiland, H. and Smith G. (1999) *Walter Benjamin: Selected Writings*. Volume 2: 1927-1934. Cambridge, Massachusetts: Belknap Press, pp. 265.

7. Jones, A.J. (2006) *Sensorium: Embodied Experience, Technology, and Contemporary Art*. Cambridge, Massachusetts: MIT Press, pp.10.

8. Jones, A.J. (2006) *Sensorium: Embodied Experience, Technology, and Contemporary Art*. Cambridge, Massachusetts: MIT Press, pp.11.

9. Glanville, R. and Muller, A. (2008) *Pask Present*. Vienna: Edition Echoraum, pp. 171.

10. Jones, A.J. (2006) *Sensorium: Embodied Experience, Technology, and Contemporary Art*. Cambridge, Massachusetts: MIT Press, pp.10.

11. Flusser, V. (2012) 'Towards a Theory of Techno-Imagination', in *Philosophy Photography 2.2*, p. 195.

12. Steiner, H. (2006) 'Brutalism Exposed: Photography and the Zoom Wave', in *Journal of Architecture Education*, vol. 59 no.3, pp.15-27.

13. Haque, U. (2007) *The Architectural relevance of Gordon Pask*. Available at: http://isites.harvard.edu/fs/docs/icb.topic983682.files/Week%2005/W05-2%20Usman%20 Haque-%20The%20Architectural%20 Relevance%20of%20Gordon%20Pask-.pdf (Accessed: 15 June 2018)

14. Levy, P. (2001) *Cyberculture*. Minneapolis: University of Minnesota Press.

15. Turkle, S. (2012) *Alone Together: Why We Expect More from Technology and Less from Each Other*. New York: Basic Books.

16. Putnam, R.D. (2001) *Bowling Alone: The Collapse and Revival of American Community*. New York: Simon & Schuster Paperback.

17. *Walker Sands Mobile Traffic Report* (2013). Available at: http://www.walkersandsdigital.com/Walker-Sands-Mobile-Traffic-Report-Q3-2013 (Accessed: 15 June 2018).

18. Kelly, J. (2014) *The lost era of the A-Z map?*. Available at: http://www.bbc.co.uk/news/magazine-26288619 (Accessed: 15 June 2018).

19. *Urban Interaction Design* (2013). Available at: http://urbanixd.eu/opinions/ (Accessed: 15 June 2018).

20. Oppenheimer, M. (2014) *Technology Is Not Driving Us Apart After All*. Available at: http://www.nytimes.com/2014/01/19/magazine/technology-is-not-driving-us-apart-after-all.html?_r=0 (Accessed: 15 June 2018)

21. Khan, O. (2011) 'Interaction Anxieties', in Shepard, M., *Sentient City: Ubiquitous Computing, Architecture, and the Future of Urban Space*. Boston: MIT Press. pp. 45-57.

22. Haque, U. (2007) 'Distinguishing Concepts: Lexicons of Interactive Art and Architecture', in L. Bullivant, *AD-Architectural Design - 4dsocial: Interactive Design Environments*. London: Wiley-Academy, Vol. 77, Issue 4, pp. 54-61.

23. Mitchell, J. W. (2005) *Placing Words, Symbols, Space, and the City*. Cambridge: The MIT Press, pp.15.

24. Lefebvre, H. (1992) *The Production of Space*. Oxford: Wiley-Blackwell.

INTEGRATING

THE INTEGRATION OF INFORMATION AND LOCATION

The media philosopher Vilem Flusser states "the development of spaces requiring separation is closely linked to the genesis of technology."[1] He might be proven wrong by the possibilities information technology presents for integration of urban space. Also the role of public spaces in cities will depend on how information technology is integrated. As globalization might have led to homogenization, apps or mobile devices that integrate information-based technology and the built environment suggest an alternative: Life online will not just be global but also local. People rely more and more on mobile devices to explore the physical world. Some may argue this development will lead to a globally homogenized urban space while others see a new opportunity to locally differentiate urban space and strengthen the identity of a physical place.

In 2000 the Global Positioning System (GPS) was made fully accessible by the U.S. government;[2] soon after the Global Information System (GIS) started to be integrated with the Internet.[3] This triggered new forms of reorganizing information. Instead of searching for information only by using keywords, information is also organized and accessed based on users' physical locations. Innovators and IT developers have often utilized location-based information in apps for smart phones and other mobile devices as main data platforms. As the Internet of Things is more integrated with location-based information, urban space might turn into a "physical moderator" of information (fig.1).

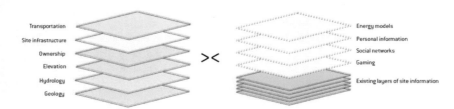

Transportation
Site infrastructure
Ownership
Elevation
Hydrology
Geology

><

Energy models
Personal information
Social networks
Gaming
Existing layers of site information

Fig.1: Integrated Spatial layers (existing and new spatial data layers).

At the same time an increasing amount of location-based information is produced every day. Sensors are used to collect real-time information from the city's infrastructure that has never been available before. Making this available to citizens by integrating real-time geospatial information in mobile devices provides citizens with a more precise understanding on how the city functions and allows citizens to use and interact with the city in a more efficient way. A rapid development of new apps is increasingly utilizing location-based real-time data. As these apps are integrated in everyday activities, one might be able to find the best route to avoid traffic, the liveliest restaurant in the area or a catalogue of events happening in close proximity. Experiencing public space through apps is a part of daily urban and social practice. As more information will become available, there is no doubt this will open up an increasing range of opportunities to rethink the way one navigates and uses urban space.

URBAN-WARE FOR THE SMART/INTELLIGENT CITY

Nowadays cities operate on database systems that provide real-time geospatial information to users. One of the main goals is efficiency of resources. Often, time is compressed to the point of avoiding any type of delay or unnecessary detour. More or less, many of us operate in these modes, guided by the real-time information delivered synchronistical to one's specific location. Many daily experiences are initiated by the data that one receives.

Real-time information generated from sensing technology and powerful computers has facilitated an insight in the city's infrastructure that has never been available before. The information collected from the city is usually owned by the government and often managed by large corporations that have the capacity to deal with large quantities of information. Making this information available for citizens by integrating real-time, geo-spatial information in their mobile devices will introduce a new understanding of the city to its citizens. Whether this new understanding will provide improved quality of life for the citizen or merely an efficiency is yet to be determined.

IBM has coined the term "Smart Cities," CISCO is talking about "Intelligent Cities" and Google is constantly providing new sets of tools from customized maps to self-driven cars to navigate the city. IBM, Cisco and other large commercially orientated developers are already tapping into this gigantic market for tools to reorganize the city. Sensors that supply live information about the cities' infrastructure or provide insights about lifestyle choices and social determinants enable the city to respond more efficiently. Large control centers such as the IBM command centers in Sao Paulo suggest a rigid top-down system. There the city's information is collected, processed and distributed. Eduardo Paes, mayor of Rio de Janeiro, demonstrated in a TED talk how to remotely control the city through gigantic IBM operation centers.

Communicating information has become a commercial opportunity for large companies like IBM but also for a great number of middle-sized businesses and small-scale entrepreneurs. Since 2013, the combination of real-time information and accessibility through smaller mobile devices has led to an explosion in the development of apps. Many of them are influencing the way people navigate the city. "Tube Exits" is an example of an app that London's citizens use to find the fastest route to a destination and the right subway car to board in order to arrive at the platform exit closest to the aimed transfer or destination (fig.2). The app, available for 99 cents, guarantees the travelers to save ten minutes on the average journey during peak hour travel. Another example is a smart phone app developed by a start-up company in San Francisco which allows taxi drivers to predict areas where rides are in high demand. At the same time the system enables cab-seekers to find others headed the same direction who want to share a taxi.

Fig.2: Tube Exits App, London.

On one hand, as more information is available for citizens, it is possible for everyone to edit and develop powerful apps. On the other hand, most apps are bought and owned by large companies such as Apple or Google. Together they are already offering more than 700,000 apps, each related to consumers who spend on average two hours per day using their apps. Open source culture suggests an urban space as a multi-author free market.

Large companies have recognized the high value of delivering real-time information in urban spaces, and these entities will determine somehow the quality of the urban experience. The citizen will in this scenario end up paying for the experience of public space by purchasing different apps, which means that large corporations not only own the market but also control the experience they provide. The only power left for the citizens is the decision to consume or not. As a result the experience of public space will emerge from consumerism. The conflicts that might emerge from this type of integration of information in urban spaces require new policies and the implementation of new regulations.

An increasing number of cities are hiring new "innovation officers" to address these issues. Open data policies are developed in cities worldwide, but technology is progressing very quickly and even the most skilled information officers will be challenged to catch up with the trends and developments of new applications and possibilities. Cities' governments will have to operate in new ways. Urban planners and designers will have to move quickly in questioning and inventing new strategies and utilizing and integrating information technology in the development and design of urban spaces.

To meet the challenge of the current trend, citizens, planners and governments not only must look for efficiency but also for quality. This requires applications that integrate information technology to contribute, for example, to the social and environmental quality of urban space. Many applications that integrate information technology and urban space are still perceived as new and experimental, but there is no doubt these apps can impact the way people navigate the city. Sooner or later some of them will reach a tipping point, and people will not just experiment with the new apps but will start to partially depend on them, as it is already happening in some places.

Information technology and new apps that help citizens navigate the city will also impact the physical urban infrastructure or the hardware of the city itself. The changing environment, characterized by climate change and urbanization, has created urgency for cities to turn into laboratories trying to squeeze out as much efficiency as possible from existing systems. Information technology integrated in urban systems, such as transportation, water supply or energy, can change the way these systems operate and relate to each other and to the citizen. In many cases systems are not connected to each other. Information technology could play a key role to integrate individual systems into a more complex urban organism.

A look at environmental developments shows how complex systems will redefine city infrastructure (fig.3). To refer to a few of those, for example, wind energy will be used to pull water out of the rainwater canal systems to irrigate plants in the parks; sun radiation will clean the air and at the same time generate energy; solar updraft towers will be integrated into building facades by using a chimney effect to suck in air at the street level that will rise up in the facade through convection; the airflow produced will have a cooling effect for pedestrians, built-in filters will clean the air from the street. The complexity of these systems is in the way they are integrated with each other to generate a systemic functioning whole. Digital and sensor technology will play a crucial

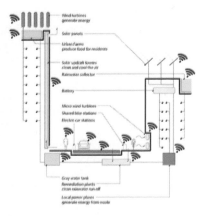

Fig.3: Integrated Urban Systems.

part in this integration, allowing buildings, streets and plazas to act as local infrastructural nodes.

Researchers at Hong Kong University, for instance, developed a modular system of wind turbines that can be used to retrofit existing building envelopes. Kinetic structures or micro wind turbines that rotate at low wind speed can generate energy from air movement caused by urban conditions and traffic; in this case information technology is used to monitor and network decentralized production of energy in urban environments. Real time information on traffic is already available for the individual. Integrating this information with a system that monitors air quality, for example, can improve the quality of urban space. Monitoring technology is also used in urban agriculture to optimize the condition of growing and processing food within urban environments. An example is TerraCycle that uses the power of social networks to recycle waste. This information can be easily integrated in communities or social platforms. Stadse Boeren[4] (Urban Farmers) is the name of an app for urban farmers in the Netherlands (fig.4). The app provides not just weather alerts but enables citizens to share what they grow in their gardens. It offers an extensive barter section, at which people trade their urban-farmed crops for those of others. It connects citizens by sharing a database of urban farming and related events and creates urban communities by providing a platform that makes it easy to get in touch with other urban farmers for exchange and to find local initiatives to meet up.[5]

The Smart City and the Intelligent City that intend to make existing infrastructures more efficient is a model that is centralized and controlled by government and large corporations. Making the city's data available to everyone allows the individual to act and participate. As people rely more on connected devices to explore the physical world, digital information will have a growing influence not only on how they move through it but also how they perceive the

Fig.4: Stadse Boeren (Urban Farmers) App and Tool-Kit.

physical realm.[6] So far it has been simply the built environment that is designed to influence the way one moves through or uses a space. Information technology will have to be integrated in the design process of urban spaces with the goal not just to increase efficiency but also quality.

THE EXPERIENCE OF DIGITAL URBANISM

Personal computers have locked people's lives into private spaces, which created a demand for people to share their private lives in public. Social media such as Facebook addressed that demand at an incredible scale. Although people need to feel comfortable trusting corporations in handling and protecting private information, social media platforms have become the most visited websites with billions of people sharing private information online. With people enjoying making their private lives public, society seems to move from one extreme, with everyone sitting at their computer at home, to the other extreme, with everyone publicly sharing every bit of private life. This will also have an effect on the transformation of physical public space. In "The Circle"[7] Dave Eggers describes a world run by a powerful Internet company. His vision draws an image of a future in which everything one does needs to be social as a consequence of being online all the time. His future is without privacy. Apps such as Dennis Crowley's Foursquare are already supporting this vision. The app grants bonus-points for discovering urban spaces and meeting new people.[8] The more social one is in public, the more points one earns.

This integrates social platforms with physical locations. This integration changes the relationships one has with public space and the way that space is occupied. The app Foursquare is available in a large variety of platforms that use social media in relation to physical space. As people rely more on connected devices to explore the physical world, digital information is changing the perception of the physical realm. A significant amount of information is already filtered in some form of digital media. The taken image is instantly shared. The user has become the medium that records physical space to provide content for the digital world. Filling the urban space with personal content could contribute to forms of revitalization. It could also integrate the private and public, digital and physical, global and local and, as a consequence, change the way one thinks about the quality of urban space.

For Bernard Tschumi, architecture starts with the event that occurs when space collides with movement, time and action. He acknowledges that the city is an intricate system of active forces in which the event establishes itself as a performance. Then the performance of urban systems is measurable by the probability of events.[9] This suggests that events can be integrated by compromising parameters of urban systems to enable the unexpected to emerge. It is, according to Tschumi, the act of manipulating parameters in the system that defines the urban event.

As a performative phenomenon, Tschumi's urban event-space calls into question both the materiality of the built-world and the immateriality of the event. The physical environment, saturated with spaces of projections and possibilities, will allow the un-thought to emerge for the unrealized and realized, the virtual and the real. In that way the performance-orientated nature of architecture redefines the architectural object, not by the mode of how it appears, but rather by the practices that take place within it.

A decade after Tschumi's *Event City*, social media platforms combined with other applications and mobile devices have dramatically changed the parameters and the form of this event. Information technology integrated as a new set of active forces triggers a wide variety of new possible events that are interactive, dynamic, temporary and light in nature. Facebook or Instagram recognizes that content on social media is not just about the event but also about the event's connection to a specific place. Therefore both are increasingly integrating location-based information in their platforms. In August 2011, Facebook introduced the option to tag photos, status updates and wall posts with location; Instagram followed in May 2013. Facebook's Nearby App, launched in 2012, offers to list friends based on proximity. Physical space being finally integrated with the content of personal information increases the opportunity for the user to search and organize location-based information.

In 2003 Bill Wasik, the editor of *Wired Magazine*, used email to bring people together in large numbers to invade stores and public spaces in New York City without any hidden commercial interest. What he later described as a social experiment named "flash mob," shows how information technology can single-handedly generate events in urban spaces and temporarily transform or activate them.

When in 2009 a Michael Jackson tribute flash mob popped up in cities all around the world, flash mob had already become a phenomenon, and social media had started to transform the use of physical public space. The different modes of social experiments in urban spaces that mobile devices make possible can quickly develop into new phenomena with large implications. Images and videos taken with a smart phone can be shared immediately with a large audience from any location. Everyone can turn into a reporter on ground. This integration of the mobile device in social platforms has contributed to movements like 'Occupy Wall Street' in 2011 and many other protests around the world. The contribution of Facebook users to catalyze a revolution at Egypt's Tahrir Square demonstrated how powerful this integration of social media with mobile devices and urban spaces can be. The political sensitiveness was also observable when in March 2014 the Turkish prime minister blocked Twitter to prevent unfavorable political information. Information technology has not only redefined Tschumi's event by integrating the physical and digital space, but it has also redefined the relation between physical spaces at local and global scales. The Internet revolution has also increasingly determined the way urban space is perceived and how its physical nature is evaluated. Instead of evaluating urban spaces based on the probability of events, urban space can be evaluated based on its experimental value for a new digital life to emerge.

For example, Pokémon Go, a location-based augmented reality game, populates public spaces with virtual objects and characters that are made visible through mobile devices. The players are asked to find and collect objects and to capture and battle creatures called Pokémons (fig.5). Invisible Playground is another platform that uses urban spaces as a gaming environment.[10] Cryptozoo demonstrates how physical urban space can gain entirely new qualities by simply overlaying a digital reality on top of it. The platform aims to get gamers into the streets by mixing computer fantasy with the real world, filling the city with strange digital creatures[11] and asking citizens to chase them throughout urban spaces. Drop Spots, another app, advertises its digital urban playground by stating that physical urban space is not exciting enough and that online applications offer improvements.[12]

Other platforms are created around different and more specific issues. Practically Green[13] is a platform that integrates game mechanics and social networks to build awareness of sustainable practices to drive measurable action locally. Mindmixer[14] is another example of an app developed in the context of open source town hall initiatives (fig.6). The app allows the government to respond and interact with inhabitants of the community. The system provides a platform for users of the system to comment and participate in activities, inviting them to generate proposals for the community and to gather and discuss ideas, problems and solutions. Platforms like Mindmixer are not only models for an interface between government and citizen but also connect and activate people around local issues

Fig.5: Pokemon Go, location-based augmented reality game.

Fig.6: Mindmixer App for open source town hall initiatives.

on a community scale. Information technology here is used to make social capital available using new communication technology.

The integration of software into the physical space suggests a model of an open city that encourages citizens to participate; although, cities will require a co-ordinated strategy to engage with digital bottom-up developments from a great number of actors. This will not only require inventing new protocols for digital tools but also challenge existing protocols used to inform the built environment.

People have been committed to new protocols offered by social media plat-forms such as Facebook, MySpace, Twitter or LinkedIn; embracing a private life lived in public will also require defining new protocols for the physical space as a flexible space with the potential to be customized by citizens' own protocols. Integrating information technology in the physical urban space will facilitate new kinds of events and redefine the idea of events themselves by changing the relationships between digital and physical, local and global, personal and public and the way in which the physical public space is perceived.

1. Flusser, V. (1999) *The Shape of Things: A Philosophy of Design*. London: Reaktion Books.

2. Official U.S. Government information about the Global Positioning System (GPS) and related topics - Selective Availability (2013). Available at: http://www.gps.gov/systems/gps/modernization/sa/ (Accessed: 15 June 2018).

3. Taylor, G. and Blewitt, G. (2006) *Intelligent Positioning: GIS-GPS Unification*. London: Wiley.

4. *Stadse Boeren Voor Leefbaarheid*. Available at: http://stadseboeren.nl/ (Accessed: 15 June 2018).

5. Wiebes, B. (2013) *Pop up City – Apps for Urban Farmers*. Available at: http://popupcity.net/apps-for-urban-farmers/ (Accessed: 15 June 2018).

6. *The new local: The physical and the digital world are becoming increasingly intertwined* (2012). Available at: http://www.economist.com/news/special-report/21564992-physical-and-digital-world-are-becoming-increasingly-intertwined-new-local (Accessed: 15 June 2018).

7. Eggers, D. (2013) *The Circle*. London: Penguin Books.

8. Warren, C. (2015) *Dennis Crowley finally has the Foursquare he always wanted*. Available at: http://mashable.com/2015/05/05/foursquare-dennis-crowley-future/#C_tohYuaMSqM (Accessed: 15 June 2018).

9. Tschumi B. (1994) *Event Cities*. Cambridge, Massachusetts: MIT Press.

10. *Invisible Playground*. Available at: http://www.invisibleplayground.com (Accessed: 15 June 2018).

11. *Cryptozoo*. Available at: http://cryptozoo.ning.com/ (Accessed: 15 June 2018).

12. De Boer, J. (2009) *Pop up City - Drop Spots: Turning Public Space Into A Secret Mailbox*. Available at: http://popupcity.net/drop-spots-turning-public-space-into-a-secret-mailbox/ (Accessed: 15 June 2018).

13. Wespire. Available at: http://www.wespire.com/ (Accessed: 15 June 2018)

14. *MindMixer- Apps for working government*. Available at: http://appsforworkinggov.devpost.com/submissions/16225-mindmixer (Accessed: 15 June 2018).

EXPANDING

THE EXPERIENCE OF EXPANDED URBAN SPACE

In 1991 Mark Weiser predicted in his essay "The Computer for the 21st Century" that computers would be embedded in everyday objects.[1] Now his prediction has become a reality also at the scale of the city. All cities infrastructural systems including energy production, water management, transportation systems, and economic and social infrastructures have been in part expanded by information technology to make them more efficient and responsive. They have been expanded to such a degree that people often rely on data feeds in operating and navigating the city. Sometimes people decide where and when they move between locations based on the prediction of traffic intensity or time left before the next bus arrives. The real-time data produced by responsive urban infrastructures can be used to gain a better understanding of how the city works and how it can be navigated or operated more efficiently. But how does this impact the experience and quality of life in cities? To assure that an increasingly sentient city is leading to an increase in quality of life, urban spaces might not only have to operate in real time but also have to engage in a continuous feedback loop of data with the users. Efficiency and immediacy of response will certainly play a role in this process.

Urban space has always been expanded and layered with information. Ancient Roman inscriptions could be seen as an early form of expanding space with information. The way digital technology is used to augment public space is fundamentally different. As information technology expands urban spaces, citizens can first access the city's information, second access information in real time, and third personalize information itself. These possibilities currently drive the interdependent developments of the hardware and devices and the software of cities.

URBAN DEVICE AND URBAN SOFTWARE

A common device that expands the experience of urban space is the smart phone. But besides functioning as a locative-media device, being a repository for a series of apps, and an instant communication tool, its potential of effectively augmenting space is yet to be fully developed. The challenge of finding new ways to expand the experience of urban space with a dynamic and responsive dimension still remains.

Bosch surround sensors

Long-range radar
Detection range ≤ 250 m, horiz. aperture angle: 30°

Night vision camera
Detection range ≤ 150 m, horiz. aperture angle: 32°

Mid-range radar front
Detection range ≤ 160 m, horiz. aperture angle: 45°

Video / stereo-video
Detection range ≤ 80 m, horiz. aperture angle: 41°

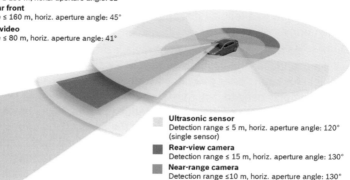

Ultrasonic sensor
Detection range ≤ 5 m, horiz. aperture angle: 120°
(single sensor)

Rear-view camera
Detection range ≤ 15 m, horiz. aperture angle: 130°

Near-range camera
Detection range ≤10 m, horiz. aperture angle: 130°

Mid-range radar rear
Detection range ≤ 100 m, horiz. aperture angle: 150°

Fig.1: Bosch mobility solutions and automated driving: surround and driver sensors system.

Digitally responsive vehicles are an example of a device one uses to physically move throughout the city (fig.1). They are equipped with a rapidly increasing number of sensors, mainly for security reasons.[2] Such vehicles can adjust their speed based on vehicles detected nearby, stop before striking an obstacle, and detect the lanes to correct driving. It can be predicted that sensor technology similar to that used in vehicles will soon be adapted for smart phones. The iPhone currently uses sensors for light and orientation and is equipped with microcontrollers that interpret data from sensors for different applications. This trend will continue; smart phones will get smarter and bristle with sensors that will provide more apps to connect the citizen to the immediate environment. Many agencies are playing important roles in this process; for instance, the healthcare industry is developing an array of body sensors for continuously monitoring one's health.[3] The gaming industry is developing new systems to track players' natural gestures in order to replace the controller that is still common in most gaming systems.[4]

Current developments lean toward decreasing the size of the device, making it invisible and integrating it with objects or with the human body. The idea's source has been around for decades in science fiction movies such as 1984's *The Terminator*. Today the Google Glass[5] experience is about to become mainstream. Macintosh and Google are developing devices that can be inserted in a contact lens. Google scientists are creating a contact lens with integrated sensors to measure blood sugar levels in tears.[6] Researchers are working to increase the resolution of wireless contact lenses by increasing the number of pixels that can be integrated into the lens. Contact lenses will evolve as smart phones have, and smart lenses will someday be as common as iPhones. Sensors and other devices are rapidly getting smaller, and eventually devices such as smart phones will be replaced by devices more integrated with the human body.

Beside the role and scale of hardware or devices another challenge lies in the potential of the development of software. Software or programming is a vehicle for agencies and actions within urban space. Augmented reality is a term commonly used to describe the possibility of extending real-life environments with the help of responsive media and interfaces that link a physical space to digital information. This means the overlay of responsive media with physical spaces could create new environmental entities that challenge the perception of space itself (fig.2). This raises three questions: 1) what can be the purpose and content of this augmented urban space? 2) what are the advantages and disadvantages of augmented urban spaces? 3) what will be the impact on the ecological, social and economic conditions of the city? Answers to these questions will provide a set of new challenges and opportunities for rethinking the urban operational model, the experience of the individual and the everyday life of cities.

Fig.2: Keiichi Matsuda, Hyper-Reality, 2013.

The notion of using media to expand urban space is not new. Artists offer a large repertory that can inspire those in charge of developing urban software or programming. The media artist Peter Weibel, for example, suggested in his early work of the 1970s that media can be used to turn the already augmented city against itself, allowing the individual to leave an imprint of his action. Banksy's graffiti suggests reframing or recycling existing space rather than creating anew. Through digital technology, everyone is empowered to leave a trace in public space. Hence it will be up to the individual citizen to sift through large amounts of data, make interpretations and decode in his or her unique way, giving birth to a generation of street computing applications.[7] These will be used to evoke curiosity, to provoke, trigger emotions and to empower citizens to be the actors in the production of public and social space.

Information technology can be a powerful tool for public engagement. In Brighton (UK), the residents of Tidy Street in collaboration with a local graffiti artist recorded on the street outside the residents' home their daily electricity use on a giant info graphic painting. The residents were given electricity meters to monitor their daily energy use and identify the type of devices using the most power.[8] In addition, open source software allowed each household to compare their energy use not only with the Brighton average but also with the national average or that in other countries. Every week the graph was updated. Residents shared strategies to save energy or simply competed with each other. The display of the information engaged people in further participation in the initiative and increased awareness regarding energy consumption, directly contributing in solving the city's challenge to save energy resources. Brighton's Tidy Street is a prototype where information technology is instrumental to involve citizens in processes that address cities' challenges (fig.3). Giving a presence to environmental information connects citizens to their physical en-

Fig.3: Brighton's Tidy Street.

URBAN MACHINES: Public Space in a Digital Culture

vironment, motivating and encouraging them to further conservation. Instead of relying on the pronouncements of city officials and information filtered through the media, the citizen is empowered to collect data, share information, draw conclusions, and engage in the development of solutions in which urban problems become the agent of the urban software.

Information technology can involve residents in the design process of the city. Visualizing information that is difficult to grasp can be a powerful tool to engage with urban planning strategies and allow residents to impact their communities. Projects such as Brighton's Tidy Street provide residents with tools and strategies to collect information for the city, report problems, vote for new proposals, provide feedback and suggest solutions. In that way, open data can turn urban space into a more democratic environment. Connecting people and places can also enrich the narrative of space by allowing urban space to be tagged with citizens' collective and individual memories, history, culture ownership and aesthetic appreciation. Allowing someone to participate in the creation of urban space, as in the case of Tidy Street's data landscape, can create a new sense of belonging to urban space itself. Information technology can be used as a critical tool in such processes.

Companies have recognized the potential of information technology to expand urban space, but one has to look closely to deeply understand the impact of some of these initiatives in the construction of the collective culture of spaces. One of the first augmented reality apps for public spaces that went live was the Metro Paris App. The app was specifically developed for the iPhone to enable citizens to visualize not only metro stations but also other points of interest by overlaying icons with the real space. Most of the "points of interest" looking down the Champs-Élysées were Häagen-Dazs, McDonald's and Starbucks. The question of who determines what one sees is obvious. The risk is that the celebrated augmented reality might turn into a commercial nightmare.

Both William S.W. Lim's book *Incomplete Urbanism*[9] and Leon Van Schaik's book *Vertical Ecoinfrastructure: The Work of T.R. Hamzah & Yeang*[10] suggest a radical expansion of urban space. Similarly and on a different scale, the performances of artist Stelarc's robotic third arm and six legs exoskeleton extend the capabilities of the human body. One of the questions at stake is how the further integration of the hardware or device and the software or programming will expand the shaping of cities in the near future.

To assure digital technology will positively reinforce public space, it will be necessary to find ways to allow citizens to be active participants and not just end users of applications provided by a few corporations. Quality of access, transparency, and usability of data are the keys that will allow citizens to experiment and participate in expanding public space through information technology:

1. Access: Internet access must be free, as a basic social right, in order to be truly public. If information technology will be fully integrated with and be part of public space, there should be no disadvantage for people who can't afford the hardware and software to access public space. The device to access the Internet will therefore have to be incorporated into the physical urban space itself. Data will have to be shared as public property, as it partially already is. Data.gov can be seen as a first attempt to make data accessible and public.

2. Transparency: The organization of data has to be made transparent. Urban software has to be open source. Code must be transparent and accessible so users can create their own spatial protocols and invent their own forms of interaction.

3. Usability: The livability of physical spaces has to be increased. People like to sit down to respond to emails. Often the only option is to leave the urban space and go to a café and become a consumer. Physical space has to be much more responsive to changing needs. Information technology allows users to quickly change from working, socializing or having a private conversation, all in the same space.

Urban space is conceptualized in very abstract urban typologies such as sidewalk, street or plaza. Information technology does not align with these conventional typologies. A Wi-Fi signal in a café, for example, provides access across physical boundaries. The way one is using digital devices in urban spaces blurs the boundaries of conventional urban types. While engaged in a phone conversation, for example, people slow down and gradually detach from physical environments looking for a place to get privacy. Depending on individual users, urban space exists at constantly changing speeds, scales and levels of presence. The types of urban spaces provided, such as sidewalks and streets, do not necessarily satisfy activities performed in public space any more. This utilitarian infrastructure was not designed for new and expanded activities. The physical urban space will have to be radically re-conceptualized.

A flexible urban space will be able to change its purpose over time. An urban space that is sometimes a market, sometimes a gathering space and sometimes used for a concert is an example for a flexible space. Information technology suggests a space that is not just flexible over time but also adaptable: It enables multiple activities happening simultaneously like listening to a concert while checking email and talking on the phone. All these activities require a space that can be temporarily stretched in terms of capacity. The physical urban space will have to be conceptualized as an adaptive, flexible space that can be responsive to multiple functions.

Rather than categorizing urban spaces in constrained forms or types, it is suggested perhaps to find new categories such as speed, situation, emotion, time, visibility, level of possible privacy or scale. Urban spaces of different speeds are spaces people experience while waiting for the bus, riding on the subway, walking or driving. Each speed changes the relationship with one's physical environment, both one's immediate surroundings such as the car or the bus and one's larger environment such as the street or the entire city. Depending on activities, people are constantly changing the scale they use to perceive public space. Someone may search for places within the immediate surrounding, another may consider the scale of the entire city while a third may look at the physical environment that is visible or at an environment made visible through augmentation apps, choosing different types of information and looking at the city in different resolutions (fig.4). Different apps are already used to learn about architecture, tourist destinations, transportation schedules, the city's history and much more.

There are many ways to expand one's perception of the city with information technology. New forms or experiences can be conceptualized as 1) adding layers of data that one perceives, 2) abstracting the physical space 3) editing the physical space by highlighting and blocking out elements or 4) intensifying urban spaces. One shall have the choice of what concept one will use or be able to create.

Fig.4: Yuichiro Takeuchi, Ken Perlin: Urban Augmentation App.

STRATEGIES TO EXPAND URBAN SPACE

Different models of expanding urban spaces have been discussed throughout the last hundred years. There are many ways information technology can change the experience of urban space again. Having identified trends in developments of concepts for digital urban devices and urban software, one can speculate on different models of urban experiences that can result from them. Five examples of such models are: digital flâneur, digital derive, digital layering, digital constructive and digital play.

Digital Flâneur: An urban landscape can be designed around behavior patterns and pre-programmed experiences. Navigating a city using Google Maps is close to this experience. The application assumes the user is interested in certain locations such as cafes, stores or restaurants. It offers these locations in a menu and makes suggestions based on other users's choices. The application will then remember the user's choice of restaurants, for example, to provide a user-specific suggestion the next time. As the number of users grows, the system analyzes the behaviors of consumers and optimizes the map accordingly.

Digital Derive: Using digital information to generate events in a city has led to a wide spectrum of projects that celebrate the notion of the spectacle to reactivate urban spaces. Paul St. George's Telectroscope uses information technology to expand the physicality of urban space. Telectroscope links a public space in London to a public space in New York by allowing for a real-time encounter between people on both sides of the Atlantic using video cameras joined through a VPN connection.

Christian Moeller crafted a robotic arm that follows passers-by and shines a perfect light circle on them at night to illuminate their space (fig.5). Both projects created public spaces that became popular meeting points where people had fun interacting with friends and strangers alike. They are successful examples in using information technology to attract people to physical spaces and populate public spaces in creating a tension between place and non-place. Inserting such interventions in urban settings subverts the predictability of urban spaces designed for the flâneur as they might be compared to the Derive's practices of the Situationist.

Digital Layering: Tying the content of the Internet to physical space enables the user to access information by wandering through physical space. What the user stumbles across determines where she or he will go next. This experience of urban space is similar to surfing the web or flipping through TV channels. This will require the reorganization of the Internet based on physical location and vice-versa. Whether what is currently rendered as info bubbles will inform the way the physical urban space is designed or not, it will definitely change the way it will be navigated.

Fig.5: Christian Moeller, Responsive robotic arm.

Digital Constructive: The alternative to an urban experience that is based on a map designed by Google can be a map that is designed by the user. Users can create their own personalized maps or use another users' maps. The user in this scenario is not passive but rather an active creator of public content. This approach reflects William J. Michell's idea explored in his book *The Cyborg Self and the Networked City* in which he stated that "The radio is one sided when it should be two sided."[11] This approach will require the development of platforms, which partially already exist, that allow people to author and share their own annotations of urban spaces.

Urban Tapestries was a research project and experimental "public authoring" platform built by Proboscis in 2003. It allowed people to 'author' the environment around them; linking photos, audio and video clips and writing to places and connecting them to other places in "threads." Inspired by the Mass Observation movement of the 1930s, it explored how people could use mobile communications to create and share everyday knowledge and experiences; building up organic, collective memories that trace and embellish different kinds of relationships across places, time and communities. Urban Tapestries enabled people to build relationships between places and to associate stories, information, pictures, sounds and videos with them – becoming active authors of the environments they inhabit (fig.6).

Digital Play: Rather than moving content from the physical world into the virtual world of the computer game, the virtual world can be moved to the real world. To make computer games more realistic, the gaming industry is mapping reality on digital avatars. Combined with sensors this also allows for a more direct interaction with the gaming environment. For example a gesture or movement in the real world can be mapped through sensors on a virtual character in real time. This process can also be reversed: Virtual environments can be mapped on real environments: Urban spaces could be temporarily expanded into playful environments, and the experience of the gaming technology could be connected to urban space.[12]

The result will be that information technology will change the meaning of physical urban space and the personal and public experience of it. New protocols, rules and behaviors are not engineered for the citizen but emerge as they interact with information technology. A new "Production of Space"[13] will be the consequence.

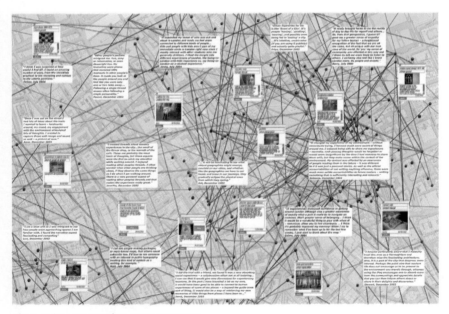

Fig.6: Urban Tapestries: Contexts Map.

1. Weiser, M. (1991) *The Computer for the 21st Century*. Scientific American, No.265(3), pp. 94-104.

2. Conner, M. (2011) *Automobile sensors may usher in self-driving cars*. Available at: http://www.edn.com/design/automotive/4368069/Automobile-sensors-may-usher-in-self-driving-cars (Accessed: 15 June 2018).

3. Pascu ,T., White, M., Beloff, N., Patoll, Z., Barker, L. (2013) *Ambient Health Monitoring: The Smartphone as a Body Sensor Network Component*. The Journal of Innovation Impact- Special Edition on Innovation in Medicine and Healthcare, Vol.6 (1), pp. 62-65.

4. Meng, R., Isenhower, J., Qin C., Nelakuditi,S. (2012). *Can Smartphon sensors enhance kinect experience?*. MobilHoc'12 Proceedings of the 13th ACM international symposium on Mobile Ad Hoc Networking and Computing, pp. 265-266.

5. Google,Google glass. Available at: http://www.google.com/glass/start/ (Accessed: 15 June 2018).

6. Poareo, J. (2014) *Google Smart Contact Lens Monitors Glucose Levels in Tears*. Available at: http://guardianlv.com/2014/01/google-smart-contact-lens-monitors-glucose-levels-in-tears/ (Accessed: 15 June 2018).

7. Robinson, R., Rittenbruch, M., Foth, M., Filonik, D., Viller S. (2012) 'Street Computing: Toward a Integrated Open Data Application Programming Interface API for Cities' in Journal of Urban Technology, n. 19(2), pp.1-23, Taylors & Francis.

8. Webb, F. (2011) *Tidy St: Shining a light on community energy efficiency*. Available at: http://www.theguardian.com/environment/blog/2011/apr/12/energy-use-households-monitor-electricity (Accessed: 15 June 2018).

9. Lim, W.S.W. (2011) *Incomplete Urbanism: A Critical Urban Strategy for Emerging Economies*. Singapore: World Scientific Publishing Company.

10. Schaik, L. (2010) *Vertical Ecoinfrastructure: The Work of T.R. Hamzah & Yeang*. Images Publishing Dist Ac.

11. Mitchell W. J. (2003) *Me++: The cyborg self and the networked city*. Boston: The MIT Press.

12. Souza e Silva, A. (2009) 'Hybrid Reality and Location-Based Gaming: Redefining Mobility and Game Spaces in Urban Environments' in SAGE-journals: Simulation Gaming, No. 40 (3), pp. 404-424.

13. Lefebvre, H. (1991) *The Production of Space. London: Blackwell Publishers*.

HACKING

HACKING URBAN SPACE

Hacking is increasingly becoming a tactic used by many spatial practitioners who operate at the intersection of digital media and urban space. Information technology that has recently expanded urban systems has initiated new opportunities to hack the city. These opportunities—if recognized by the individual citizen—provide a powerful tool for change through questioning, altering, or subverting an existing system. Rather than waiting for city officials or private developers to take action, hacking could empower every citizen to participate in the construction of public space. As hacking has recently become a tool for a number of projects and organizations that speculate on the development of public spaces such as "Civic Hackathons," one can assume that hacking will increasingly develop as a strategy to empower the individual citizen to intervene in urban environments. This suggests a new form for the citizen to navigate the city, to understand it, and to interact with it in new and meaningful ways. This will not only change the urban environment but also challenge urban planners, architects and city officials to rethink the current instruments and methods used to shape cities.

The term "hacking" became popular in the digital subculture of the 1960s. The motive of the hacker is generally understood as gaining unauthorized access to a computer system to destroy data, to access information for personal purposes, or to use the gained access to distribute messages with social, ideological, or political content. Such actions are often small manipulations of complex systems leading to great consequences. In the 1980s the term was popularized by movies such as *Blade Runner*, *Tron* or *WarGames*. In *WarGames for example*, high school student David first hacks into the school district's computer system to change his grade and later accidentally hacks into an automated missile strike system at NORAD that almost causes a nuclear war.

Movies like *WarGames* have contributed considerably to a common understanding of hacking as a criminal or highly dangerous action usually executed by an individual and targeted against a large and powerful entity. Since 2001 the War on Terror has changed the perception of hacking. Now large and powerful government organizations commonly hack into the space of an individual. The U.S. National Security Agency hacks into security systems, the Internet and telephone systems all over the world. Similar activities are executed by other powerful nations that sometimes control activities within these systems. "Public space" is a victim of the method. In many places, closed-circuit television cameras hack into the public space and follow, record, and analyze every step of the individual occupying it. It is the individual hero, such as Jason Bourne,[1] who knows how to navigate this new situation and stay invisible. Who should hack, and for what reason are the issues at stake. The topic of this chapter refers to how the act of hacking can support and transform public space.

The term hacking does not need to carry a negative connotation. Indeed, computer programmers often use the term in a positive way. Exploratory programming workshops—called Hackthons—team up software developers with communities to develop open source solutions to problems using publicly released data. Looking at the act of hacking in the context of public space, it is suggested to leave the immediate associations with the term behind in order to reintroduce hacking as a tactic in a broader sense. In that way, hacking means gaining access to a system in order to manipulate it. This definition creates a framework for a body of work from individuals, artists and organizations operating within these new parameters.

Peter Weibel's manipulation of visual urban systems and language in the 1960s can be considered as an early form of hacking into public urban spaces (fig.1). In one memorable instance, he held up the words "are lying" next to the "police" sign on a station, intentionally trying to provoke a reaction from passers-by. Such very minimal events can be understood as a precursor of actions that make alternatives evident by subversion. In the 1970s many artists and architects followed Weibel's vision in transcending the gallery space and hacking into public space.

Haus-Rucker-Co, a group of Viennese artists, is another example of positive hacking. Their designs for inflatable structures, prosthetic devices and interventions to hack into public spaces were prototypes installed in an urban space to promote social change, an experiential theory of space and the destruction of public space and private space for a new environment. Their temporary installations were called "provisional structures" to hide them within the legal system. Their ideas—often seemingly impossible—drew them to use materials considered strange, new, and unusual at the time. One of these provisional structures was a huge Perspex ball that was cantilevered from the window of a 19th century building (fig.2). The Perspex ball extended the private space of the building into the public space, forming an almost personal oasis suspended 10 meters above the ground.

Other projects by Haus-Rucker-Co included "Environmental Transformer" (fig.3) and "Mind Expanders" (fig.4). Mind Expanders enabled people to sit together in a public space and at the same time being completely isolated

Fig.1: Peter Weibel, Polizei lügt, Vienna.

Fig.2: Haus-Rucker-Co, Oasis, Vienna.

from the outside world. Environmental Transformer, a bottle green Perspex double bubble head piece with its own power pack was made for people to wear in public spaces. This head piece not only provided people with a fly's eye perspective on the space they occupied, but it was also designed to completely change the relationship between the wearer and his surrounding environment. In the context of the current debate on how digital information technology might change the experience of public space, it is no wonder 1960s artists such as Haus-Rucker-Co have been rediscovered and celebrated in contemporary exhibitions. Today artists are manipulating digital information technology systems to bring our attention to our everyday accepted norms in public space. The artists—often performing a small change in the system—cause a large impact on the perception of public space.

Fig.3: Haus-Rucker-Co, Environment Transformers, Vienna.

Fig.4: Haus-Rucker-Co, Mind Expanders,
Vienna.

In June 2007 the art group Ztohoven hacked a camera used for a live broadcast on CT2 of Czech Television.[2] Ztohoven piped a video of a nuclear explosion and a mushroom cloud onto a live panoramic view of the Krkonose Mountains, a well-known tourist destination. The project caused calls from a worried TV audience and led to legal action against the artists. Charged with public gullibility, scaremongering, and spreading false information, the artists faced prison sentences of up to three years.[3] After the judge dismissed charges against them citing "public amusement rather than public unrest,"[4] Ztohoven received a prize for Media Reality from the National Gallery of Prague. Its president Milan Knizak commented: "Ztohoven left the gallery space entering the public space where they provoke society."[5] The project shows that even the slightest intrusion can appeal to the intellect of citizens as reminder that there is a difference between reality and mediated reality and that there is a need to question the trueness and credibility of media.

Ztohoven recently hacked into Prague's urban infrastructure, replacing 48 Ampelmännchen (symbols of a standing or walking person commonly used as pedestrian signals) with their own figures shown in situations such as drinking, urinating, or being hanged (fig.5).[6] The artist was drawing attention to the way pedestrians unquestioningly obey these figures as they navigate the city streets daily. The artist's new variations of Ampelmännchen could be only seen for one day before the city changed them back. The project was experienced by the public as great fun. The artist was sentenced to one month in prison.

Even the slightest manipulation of public space can put the individual in conflict with the legal system. Artists have therefore developed different attitudes about how to navigate legal boundaries. This is for example demonstrated by the group The Surveillance Camera Players. By performing, pointing and even appearing to pray to surveillance cameras in public spaces,[7] the group critiques the authority that spies on people in public space but manages to do so without breaking any laws. In contrast, public space hackers who play the games Camover and Killcap in Germany clearly run afoul of the law and would face sanctions if caught destroying government-placed cameras in public spaces. These gamers film themselves destroying the spy cams and upload their footage onto a website where they earn points for each destructive act.

After 2010, hacking urban spaces, usually for political or social reasons, became increasingly widespread as an artistic practice. In that year "Hacking the City" was the title of a project in Essen, Germany, to celebrate the city's election as the European Capital of Culture. The intention of the hacking endeavor was to react to the city's changing structures of public space, mobility, and communication by reprogramming and alienating urban spaces. One artist who contributed to the project was Peter Bux[8] who staged an apartment move by

Fig.5: Ztohoven Guerrilla Artists, Hacking Traffic Lights, Prague.

piling up boxes and furniture on a sidewalk that over time grew into walls and blocked traffic. Other contributions included a guerrilla gardening project by Richard Reynolds[9] and toilet seats displayed in public spaces by Stefanie Trojan.[10]

All these projects temporarily physically disrupted urban systems in the city of Essen, which raises the question of whether hacking can lead to long-term change. In 2007, the artist Natalie Jeremijenko transformed the "dead" street spaces around fire hydrants into tiny parks to absorb road-born pollutants and storm-water runoff. The parks were designed to allow access for the firefighters, making them legally possible and suggesting that the interventions have the potential to permanently change streetscapes.

Other artists see themself as facilitators for the citizen to act. Architect Santiago Cirugeda's interventions hack into the city's hardware by subverting regulations and laws to improve the everyday urban space. In his call for action titled "Building yourself an urban reserve" citizens are asked to review, reinterpret and reuse the Seville General Urban Zoning Plan Ordinance that governs the placement of temporary scaffolding (fig.6). Citizens then are asked to use the regulations to their advantage in expanding their buildings

Fig.6: S. Cirugeda, Building yourself an Urban Reserve, Seville.

using scaffolding installed on the public space in front of their property. The intervention, a temporary room connected to the houses' interior has to be accessible from the public space of the street, as it is required by the law. His practice of appropriation and occupation of urban space understands people as the creators of urban space, questioning the notion of authorship and control. Most of his projects are open source. Cirugeda's "Urban Prescriptions"[11] website offers a user's manual that enables others to replicate his systems.

A growing integration of wireless tools and infrastructure into the everyday life of a city can lead to an increase of possibilities to hack into these networks by the individual citizen, carrying fundamental consequences for the public realm. To enable individuals to hack into the urban space to appropriate, manipulate, revaluate, and reinvent it, will require making public space hack-able for everyone. Only this open source strategy will enact hacking as instrument to improve urban space.

OPEN DEVICE AND CODE

The urban environment is both a generator of data and the product of urban information ecology. Transparency and accessibility of urban information can lead to catalyze strategies for open source urbanism. Digital devices can be used to access cities' information ecology but can also become a tool for hacking actions.

Pervasive networks allow the city to become the generator of information. Sensor systems applied to the urban scale—able for example to monitor weather, traffic, or temperature—function as independent systems that operate relatively free from human interaction. This artificial eco-system monitors, processes, and manages information. The data produced by it can be used to inform other systems in real time. The collection and exchange of data is based upon the communication from device to device. As such, the urban information ecology creates an overarching network that can, in theory, operate independently from human interaction.

Organized in what Marshall McLuhan calls a "galaxy of machines," this "electrical environment," forms an extended nervous system that is both invisible and pervasive.[12] He refers to this hidden and unseen artificial eco-system as an "environment of services." The future will be a "world of connected machines"[13] that function autonomously, "talking to other machines on behalf of people." Such a degree of automation effectively enables machines to "read and write by themselves."[14] This condition of simultaneity, instantaneity, intelligence and interrelation resonates with Marcel Mauss's definition of "savage telepathy," a scene in which "the whole social body comes alive with the same movement." The play of instant machine correspondence suggests an "intelligence" of exchange where anticipation and event coalesce in the savage communication of machines.[15] These continuous communications facilitate the city's ecologies that regulate its rhythm.

Information technology devices for surveillance or automation are assembled into an intelligent communication ecology, a "new sensorium" regulated by the communication of devices, without taking into account human intervention. The city continuously sends back signals to itself to regulate itself. The temperature is 70 degrees, the wind is blowing from east, the traffic is flowing slowly, and the noise level is too high therefore the traffic will be rerouted. All this data quantified, measured, and integrated into a system that mutually influences itself allows the city to be interpreted as a self-regulating organism. The senses and the brain of the organism are in a continuous dialogue even without conscious activity. Phenomena are the type of data that feed the device, and the device sends signals to regulate the city's internal system. As Paul Virilio argues, the electronic communication changed the physical fabric of the city. The surge of communication through the electronic ether gives rise to a city devoid of spatial dimensions but inscribed in the singular temporality of an instantaneous diffusion. "The city is overexposed: it exists all at once."[16]

Knowing that the temperature is 70 degrees, the wind is blowing from east, the traffic is flowing slowly, and the noise level is too high, an additional challenge is to use this information to increase spatial quality and not just performance of function.[17] The city functions can adapt to information technology much faster than the space. Its process of instantaneity reconfigures the relation between space and time. Stephen Graham asks us to "imagine the 'real-time' city" so one can account for the ways in which telecommunications reconfigure our notions of urban space and time.[18] Sensors and mobile devices, and their machine-to-machine communication, reconfigure urban ecologies. The city is at once observable as an artifact, an object of data and surrounds us as a space of potentially limitless data production.[19] The city has a mind, a processor that can always be inquired for an answer.

Increasingly instant and automated, urban space circulates through the transitory and traffic monitored circuits of the web cameras, surveillance systems, timers and traffic monitors. Citizens can certainly benefit from this real-time data. A simple example is the display of time left for the next bus to arrive. Citizens are usually seen as the end user in this scenario unless they become active agents, gain access to the city's data and reconstruct what defines public space. In this scenario, hacking strategies become an operational tool to act and to transform the information urban eco-system.

OPEN SOURCE CITY

Making systems of hardware and software more accessible in recent years led to citizen initiatives transforming urban space. Open source concepts allowed for initiatives to realize urban gardens, community spaces, shared Wi-Fi (WLAN) zones, or projects concerned about environmental monitoring. "Open source urbanism" develops where citizens gain access to the information that shapes urban space and turns them into agents. Open source systems provide the individual with new possibilities to hack and manipulate those systems to directly inform the urban space.

The idea of open source is associated with free computer programs that can be shared, adapted, and further developed by any user.[20] Applying this idea to urban space means that all systems that make up urban space are accessible to everyone and are connected to all other systems. It is a concept of horizontality and distributed network. Saskia Sassen argues that the city is in a state of incompleteness and that the concept of "intelligent cities," as it is only taking into account hardware, will be soon obsolete. Open source urbanism that is grounded in the software of social practices allows for bottom-up interventions that will continue to emerge. Every day opportunities for events will occur by the individual instrumentation of information technology of urban space. This increase of individual agency will shift our attention from the global to the local network. To do that, it will be important to recognize that technology is not only about the device but also about the instrumentation of the device.

As information technology is pervasive and ubiquitous, local technology activists experiment with the construction of new tools to rethink the relationship between citizens, their governments, and communities. These actions in an open source urban environment of civic technology build on our already networked culture and promote a more efficient system of collaboration between entities that produce the city. The opportunity for an individual to make an app that has a large impact has exploded; of course, a financial motivation drives this industry as well.[21] This growth has also created new communities of citizens, software developers, and entrepreneurs who meet for Hackathon workshops.

In June 2013 during the first National Day of Civic Hacking[22] more than 90 Hackathon workshops were organized simultaneously across the United States with the goal to motivate citizens to contribute in changing their communities through open source, open data, entrepreneurship and code development. The event brought together citizens, software developers and entrepreneurs from all over the nation to collaboratively create, build and invent new solutions using publicly released data, codes and technology to solve challenges related to individual neighborhoods, cities, states and the country. In each city the event addressed different issues depending on local needs. Projects included apps to predict commute times and apps that help users make financial decisions. Other apps assist urban farmers in enhancing the experience of farmers markets or create remote and local user interfaces for data of plants. During the events, expert technologists encouraged anybody interested to use publicly available data sets to imagine solutions that benefit the everyday life of the citizens. During the Hackathon, the White House posted on its blog: "This is an opportunity for citizens in every town and citizens across the nation to roll up their sleeves, get involved, and work together to improve our society by cultivating an ecosystem for innovation and change."[23] The challenge set up by Hackathons is to liberate and democratize open data to support problem solving in every community.

This goal includes the vision of increasing collaboration and facilitates methods of sharing. Code for America, an organization involved in the development of technologies that change the conversation between citizens and government, proposed the following "10 Ways Civic Hacking is good for Cities." The goals were to create space for innovation, engage digital citizens in the process of governance and creative problem solving, spur economic opportunity, provide insight into government decision making, enable community service by technology, teach important new tech skills, create a broad network of civic hackers, help citizens serve themselves, help government manage expectations around technology and connect technology and non-technology groups together. By open source code these suggestions propose a vision of a city that is able to create a strong connection between the citizens, the government and future urban scenarios. Anthony Townsend states in his book *Smart Cities: Big Data, Civic Hackers, and the Quest for a New Utopia*[24] that the current world is defined by urbanization and digital ubiquity, where mobile broadband connections outnumber fixed ones, machines dominate a new Internet of Things, and more people live in cities than in the countryside. Cities worldwide are deploying technology to address both the timeless challenges of government and the mounting problems posed by human settlements of previously unimaginable size and complexity. Anthony Townsend talks about the strong potential of open source data as an option for future urban management. The act of hacking is using open source data to move towards more efficient, more resilient and more democratic cities. The "Smart Citizen" in this scenario is the empowered citizen who proposes solutions rather than waiting for the government to resolve problems.

Air Quality Egg,[25] a community-led network of sensors, is just one of many examples of civic empowerment (fig.7). Using the web and a sensor system-kit, anyone can report on the air quality outside of their home. In this example, individual citizens are participating in the production of global data and at the same time creating a debate about it.

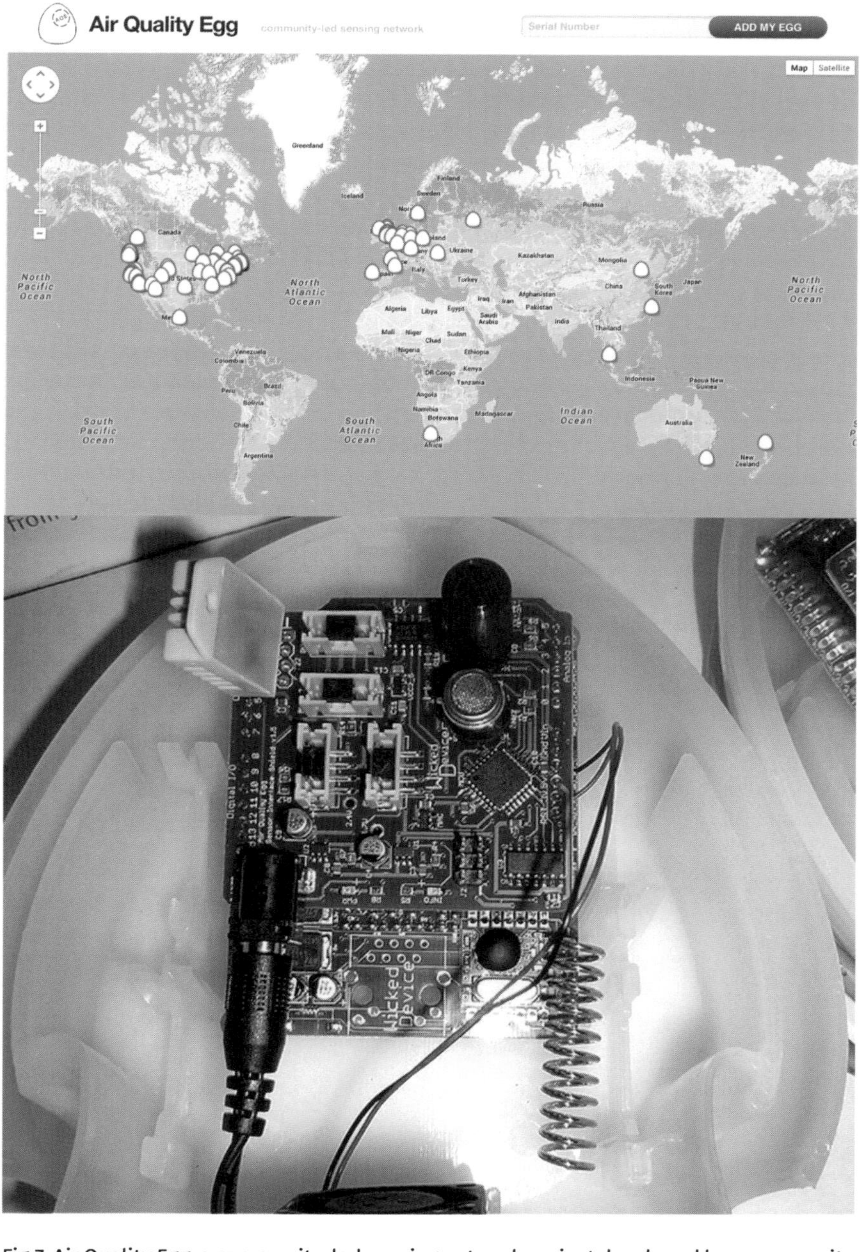

Fig.7: Air Quality Egg: a community-led sensing network project developed by a community effort born out of groups from the Internet of Things Meetups in NYC and Amsterdam.

Another project, called "Smart Citizen" (fig.8) is a "do-it-yourself kit" that enables citizens to be part of mass environmental monitoring.[26] In a similar project, a guerrilla group of citizen-scientists installed sensors in local sewers in New York City to alert citizens when storm water runoff overwhelms the system, dumping waste into local waterways.

These types of projects, leveraging from democratized technology and open data, enable the individual citizen to step forward and deploy solutions for improving communities. In regard to the process and time for these actions to take place, Anthony Townsend says, "We need a lot more sustained energy, cohesion and leadership in the civic tech movement for it to have a real long-term impact, and to deliver the innovation potential that is there. Kickstarter projects are a good place to start, but what gets me excited is seeing industrial giants like Intel embrace Arduino, an open-source hardware and software. They see the future in an Internet of Things that people build themselves, and parallels to the PC revolution in the 1970s." The key will be to dramatically increase the number of hackers from a small group of artists to the larger citizenry. "I really think it is the key to a more bottom-up, urban design-driven vision of a smart city—not as a place enabled by big smart infrastructure, but one that accumulates organically from thousands and millions of tiny little installations."[27]

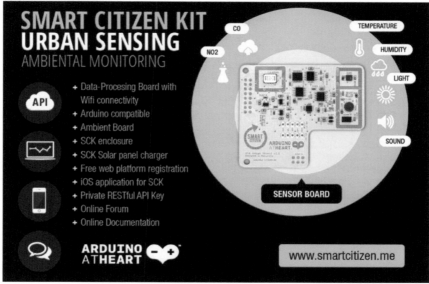

Fig.8: Smart Citizen_ Do-it-yourself Kit project developed at IaaC, Institute for Advanced Architecture of Catalonia, Barcelona.

1. The protagonist of a series of novels by Robert Ludlum that in 2002 was adapted into a feature film.

2. Ztohoven, *The Media Reality*. Available at: http://www.ztohoven.com/?page_id=45&lang=en (Accessed: 15 June 2018).

3. Kimmelman, M. (2008) *That Mushroom Cloud? They're Just Svejking Around*. Available at: http://www.nytimes.com/2008/01/24/arts/design/24abroad.html?pagewanted=all&_r=1& (Accessed: 15 June 2018).

4. Van Gelder, L. (2008) *Hacker Artists Cleared in Czech TV Stunt*. Available at: http://www.nytimes.com/2008/03/27/arts/27arts-HACKERARTIST_BRF.html (Accessed: 15 June 2018).

5. ArtLaboratoryBerlin (2009) *Ztohoven - Media Reality*. Available at: http://artlaboratory-berlin.org/html/de-ausstellung-14.htm (Accessed: 15 June 2018).

6. Schmidt, V. *(2011) Urinierende Ampelmännchen – Künstler verurteilt*. Available at: http://www.welt.de/vermischtes/article13753416/Urinierende-Ampelmaennchen-Kuenstler-verurteilt.html (Accessed: 15 June 2018).

7. *The Surveillance Camera Players* (2006). Available at: http://www.notbored.org/the-scp.html (Accessed: 15 June 2018).

8. Bux, P. (2010) *Achtung Umzug!*. Available at: http://www.hacking-the-city.org/artists-and-projects/peter-bux.html (Accessed: 15 June 2018).

9. *Richard Reynolds* (2010). Available at: http://www.hacking-the-city.org/artists-and-projects/richard-reynolds.html (Accessed: 15 June 2018).

10. *Stefanie Trojan* (2010). Available at: http://www.hacking-the-city.org/artists-and-projects/stefanie-trojan.html (Accessed: 15 June 2018).

11. Cirugeda, S. (2007) *Recetas Urbanas*. Available at: http://www.recetasurbanas.net/v3/index.php/en/ (Accessed: 15 June 2018).

12. McLuhan, M. (1964). *Understanding Media: The Extensions of Man*, Boston: The MIT Press.

13. Gabrys, J. (2010) 'Telepathically Human' (Paul Saffo cited Johnson 2000) in *Circulation and the City: Essays on Urban Culture (The Culture of Cities)*, McGill Queens University Press.

14. Gabrys, J. (2010) 'Telepathically Human' (Kittler, 1997) in *Circulation and the City: Essays on Urban Culture (The Culture of Cities)*, McGill Queens University Press.

15. Gabrys, J. (2010) "Telepathically Human" in *Circulation and the City: Essays on Urban Culture (The Culture of Cities)*, McGill Queens University Press.

16. Virilio, P. (2003) *Art and Fear*, London: Continuum.

17. Riether, G. (2011) 'Towards digitally integrated urban places' in *Kybernates*, No. 40 (7/8), pp. 1129-1136.

18. Graham, S. (2005) 'Imagining the real-time city: Telecommunications, urban paradigms and the future of cities" in *Imagining Cities: Scripts, Signs, Memory* edited by Sallie Westwood and John Williams, London: Routledge.

19. Gabrys, J. (2010) 'Telepathically Human' in *Circulation and the City: Essays on Urban Culture (The Culture of Cities)*, McGill Queens University Press.

20. Artibise, Y. (2010) *Open Source Urbanism: Where Data meets Urban Form*. Available at: http://yuriartibise.com/blog/open-source-urbanism-where-data-meets-urban-form/ (Accessed: 15 January 2017).

21. Leckart, S. (2012) *Th Hackathon is On: Pitching and Programming the Next Killer App*. Available at: http://www.wired.com/magazine/2012/02/ff_hackathons/all/1 (Accessed: 15 June 2018).

22. *National Day of Civic Hacking* (2016). Available at: http://hackforchange.org/ (Accessed: 15 June 2018).

23. Knell, N. (2013) *White House Drums Up Support for National Hacking Event*. Available at: http://www.govtech.com/e-government/White-House-Drums-Up-Support-for-National-Hacking-Event.html (Accessed: 15 June 2018).

24. Townsend, A. (2013) *Smart Cities: Big Data, Civic Hackers, and the Quest for a New Utopia*, New York: W. W. Norton & Company.

25. *Air Quality Egg: Community-led sensing network*. Available at: http://airqualityegg.com (Accessed: 15 June 2018).

26. *Smart Citizen Platform*. Available at: https://smartcitizen.me/ (Accessed: 15 June 2018).

27. Dele, B. (2013) Anthony Townsend on *Hacking Into 'Smart Cities'*. Available at: http://nextcity.org/civic-tech/entry/interview-anthony-townsend-on-hacking-into-smart-cities.
(Accessed: 15 June 2018).

NETWORKING

NETWORK STRUCTURES AND EMERGING URBAN FORMS

In the Internet age networked structures have become the organizational model of cultural and technological production. A network is an abstract organizational model in its broadest sense concerned only with the structure of relationships between things, be they objects or information.[1] Social networks that resulted from technical infrastructure have generated new categories of public commons. In the last twenty years the increasing emergence of telecommunication networks and the understanding of network structure in relation to space have situated network forms within the discussion of future urban environments. Many questions arise by thinking about how these networks inevitably affect almost all of our daily activities.

Manuel Castells states in his book *The Rise of Network Society*: "The network society itself is, in fact, the social structure which is characteristic of what people had been calling for years the information society or post-industrial society."[2] The relationship between networks and contemporary society underlines the importance as well as the opportunities given by forms of networks to establish conditions for mutated concepts of social-cultural space.

In this context network society is understood as defined by Jan Van Dijk as "a society in which a combination of social and media networks shape its prime mode of organization and most important structures at all levels (individual, organizational and societal)."[3] Similar positions are also offered in James Martin's book *The Wired Society: A challenge for tomorrow*.[4] The decreasing supremacy of the street or the plaza as the main meeting point and space has led to a development of what William J. Mitchell calls "electronic agoras." He argues that the worldwide computer network—the electronic agora—subverts, displaces, and radically redefines our notions of gathering, place, community and public life.

The network has a fundamentally different physical structure, and it operates under quite different rules from those that organize the action in the public places of traditional life. It will play a crucial role in the 21st century urbanity just as the centrally located, spatially bounded, architecturally celebrated agora played in the life of the Greek polis.[5]

Michael Batty and Andrew Hudson-Smith argue in their essay "The Liquid City" that in the 19th century energy was the catalyst to expand cities and the connector between physical territories that were otherwise isolated. The shift from energy to information, from "atoms to bits" as eloquently phrased by Nicholas Negroponte, is changing cities in ways that are not visible at first sight. Globally the effect of such communication is that cities are starting to merge into one another, if not physically, then digitally.[6]

Two questions arise: 1) How do information-based environments affect traditional urban typologies? 2) How are designers able to shape the agency of networks? The direction for answering the first questions is: If in the past, until the end of the 19th century, cities were usually physically connected and able to expand and conquer new territories, they are now witnessing the shirking and replacement of some obsolete urban functions with others that enable the expansion of the digital network. This network expands by capturing the various flows of exchange in the city. The city begins to grow from a series of nodes that are all connected reversing the mono-centric condition to the polycentric form. The future is likely to reflect, through the physical form, the many levels of complexities and opportunities of the network itself, where overlapped, multi-layered, simultaneous and metabolic conditions will be the operative terms used to re-think future urban development.

An answer for the second question can be found in Lee Stickells' essay "Flow Urbanism": "The interest in flows can be positioned within a wider discussion regarding the nature of the contemporary city and the emerging tensions between its fragmenting physical fabric and multiplying electronic socio-economic networks."[7] Within the logic of space developed through flows as the physical entity of network forms, Paul Virilio argues for a transfiguration of architectural materiality given an increased level of connectivity. "The old agglomeration disappears in the intense acceleration of telecommunications, in order to give rise to a new type of concentration: the concentration of domiciliation without domiciles, in which property boundaries, walls and fences no longer signify the permanent physical obstacles."[8]

As the city is continuing to spread into physical and non-physical nodes and links, this diffuse vast plane can be described as Rem Koolhaas' generic city. Koolhaas describes the contemporary process of urbanization: "if there is to be a "new urbanism" it will not be based on the twin fantasies of order and omnipotence; it will be the staging of uncertainty; it will no longer be concerned with the arrangement of more or less permanent objects but with the irrigation of territories with potential; it will no longer aim for stable configurations but for the creation of enabling fields that accommodate processes that refuse to be crystallized into definite form. It will no longer be obsessed with the city but with the manipulation of infrastructure for endless intensifications and diversifications, shortcuts and redistributions—the reinvention of psychological space."[9]

Koolhaas envisioned urban forms moving into a space that embodies the flux and the formal qualities of the infrastructural network. Smooth space envisioned by Gilles Deleuze and Pierre-Félix Guattari embodies the agency of the network dissolving urban typologies conceived as objects. "The smooth is the continuous variation, continuous development of form and the points are subordinated to the trajectory."[10] Flat space, horizontal and distributed are the spaces of encounters. The taxonomy of the city as square, street, park, and so forth is subverted by the logic of non-hierarchical forms. Here is what the network physically could embody, because network topologies engender forms of horizontal encounters and node densification rather than supremacies of spaces. The city is determined by diffuse systems of relationships overwriting hierarchies.

The spatial distribution of networks is reflected in the pattern of city-growth that mimics network morphologies. Polycentric and distributed cities are evolving into constellations of nodes connected by both high-speed transportations and digital networks. New socio-spatial realities emerge encompassing the metropolitan and the global scale. As a result, governors and city mayors, educational institutions, e-entrepreneurs, the information technology industry, community developers, planners and urban designers among others have come together to reinvent locales as more livable, sustainable and vibrant digitally connected communities. The rallying cry of these coalitions is often a denunciation of urban sprawl and its consequences, including central city decline, lack of affordable housing, long commutes, traffic gridlock, fast-disappearing open space, environmental pollution, dependency on cars and mass-produced and boring development patterns.[11]

THE (SOCIAL) PRODUCTION OF SPACE THROUGH THE NETWORKED PUBLIC

Buckminster Fuller and Marshal McLuhan imagined how information technology might impact architecture and urban space long before the arrival of the Internet. Both imagined the consequences of information networks on the built environment. In 1938 Fuller suggested a worldwide energy network (fig.1) and a housing project based on the telephone network.[12] In 1962, McLuhan coined the term "global village" and predicted in his book *The Gutenberg Galaxy: The Making of Typographic Man*[13] the Internet thirty years before its arrival. The argument was that the digital network would be a catalyst to create a worldwide community. By the 1990s the Internet became global and started to be incorporated into general daily life. The invisible infrastructure of digital networks has been realized and is constantly feeding us with new information. The focus has been partially shifted from the physical city to the immaterial infrastructure attributing to the information network the ability to instigate new forms of social and cultural experiences.

The boundaries between invisible digital networks and the physical city have blurred. An understanding of the urban environment as a network is suggesting new hybrid conditions that are both inter-scalar and metabolic. Over the past few years the collapsing of such physical networks with non-physical has led to a new condition, an urban experience defined by hybrid networks that are both, partly physical and digital (fig.2).

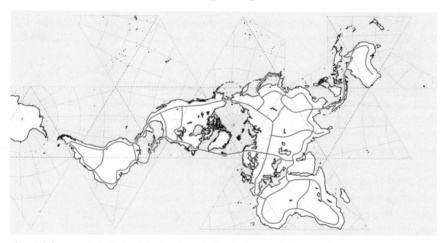

Fig.1: Vision of global electrical network, Buckminster Fuller, 1938.

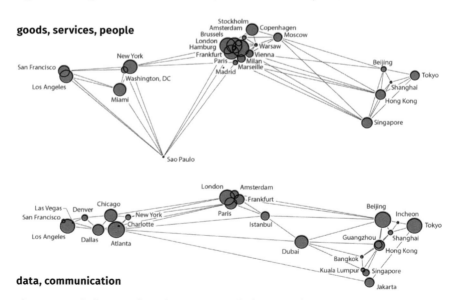

Fig.2: Network diagrams based on a 2015 analysis by Mc Kinsey Global Institute.

These networks are transferring parts of the physical urban experience to inform new digital layers of information. Such networks have become active agents in the experience of everyday life and an important parameter for any form of spatial practice. Daily activities are informed and influenced by them as people move towards a new type of publicness that has new needs and modes of physical and non-physical encounter, a public that will enable the continuous mediation between the digital and the physical space. Architects, urban planners and designers should engage these dynamics through a design perspective as the physical city is re-imagined as a sentient being in a continuous state of flux.

Network technologies might contribute to the dispersal of private activities throughout public space while they are also able to promote and stimulate collaborative public forms. Free hotspots, currently implemented in many cities throughout the urban fabric, are one example. Such public open nodes dispersed throughout the city will be useful on multiple levels such as enhancing city management, public safety and stimulating economic growth and providing new platforms for social interactions. At the same time, the mobile device has led to a personalization of public space.

Bike-sharing systems demonstrate how citizens interact with hybrid networks that are constructed from both the digital and physical. One of the largest networks of bike-sharing stations is in Paris; it is called *Velib* (fig.3). By using smart phones people can search for the closest station and check for

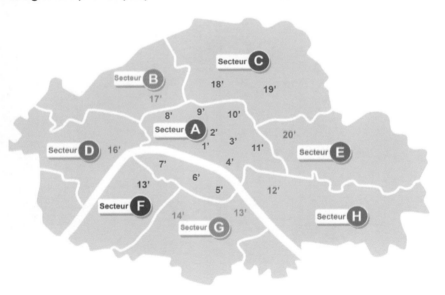

Fig.3: Map of Velib, a bike sharing system of Paris.

URBAN MACHINES: Public Space in a Digital Culture

available bikes to be taken for biking through the city and—at the same time— the system is creating new bike-sharing communities. This type of system is quite common and now shared by major cities throughout the world.

MapMyRun is another prototypical example of such a system. It is a mobile app that enables citizens to track their running route, time, distance, speed, pace and calories in real-time for their urban fitness activities using GPS integrated in their mobile device. It suggests a route to run but also invites individual runners to share their routes. The individual is not just following a map but is becoming a map-maker for others; the sharing of information instantly creates new communities and influences how the physical space is experienced.

As an individual device the "computer" is going to be superfluous, in its place all of the surrounding and everyday objects will be equipped with digital technologies.[14] Cities will become an ecosystem of mutually communicating objects, buildings and virtual environments.[15] These ecosystems will generate new communities and a new form of public. Objects, people and places will become increasingly connected through network forms while becoming active agents in the process of urban production.

LOCAL - GLOBAL / ACTIONS / AGENCIES

The sentient city emerges from the continuous hybridization and the layering and overlapping of physical and digital systems. The word "sentient" implies a system that is conscious or aware and able to respond.[16] Applying this concept to urban scale, one can envision a model that uses digital networks to integrate physical space and technology at a social level to produce a new system of relationships that encompasses shared knowledge, collective actions and coordinated interactions between individual actors and the collective.

The systems that make up the sentient city are able to learn and will become "smart." Carlo Ratti, director of the MIT-SenseLab, describes in an interview with *Wired* magazine how citizens will become the vehicle of such networks: "By receiving real-time information, appropriately visualized and disseminated, citizens themselves can become distributed intelligent actuators, who pursue their individual interests in co-operation and competition with others, and thus become prime actors on the urban scene. Processing urban information captured in real time and making it publicly accessible can enable people to make better decisions about the use of urban resources, mobility and social interaction (fig.4). This feedback loop of digital sensing and processing can begin to influence various complex and dynamic aspects of the city, improving the economic, social and environmental sustainability of the places we inhabit."[17]

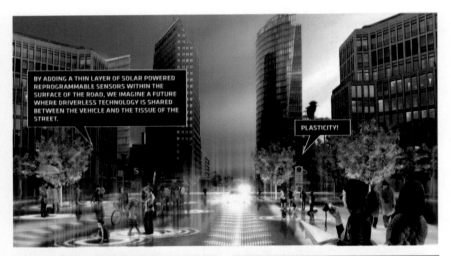

Fig.4: Audi Urban Future, BIG – Bjarke Ingels Group.

Over the last years the worldwide web has been the terrain for public debate and collective organization. Through the Internet, numerous protests and movements originated and instigated by the open network enabled individuals to organize themselves as groups to take action and to subvert top-down rules. Open processes of collective self-organization have increased; the agent being the network, the actuator is the individual citizen. The 2012-13 Egyptian protests are one of many examples. As the military closed Tahrir Square from demonstrations social media pages such as "We Are All Khaleed Said," with more than 1.6 million followers were used to organize protests and campaigns elsewhere in Cairo with thousands of people participating. Another example is the increasing number of non-political events such as flash mobs performances organized through social media. Masses engaged at that very large scale through social media networks are more and more affecting the meaning of public space.

Bruno Latour argues in his book *Reassembling the Social: An introduction to Actor-Network Theory*[18] that the network has the capacity to perform. The network of a sentient city is made up of humans, objects and digital technology in which the actuators and actors are not only humans but also technologies. Latour claims that not only humans have agencies in the creation of the urban spaces, but also technologies. In this framework, both human and non-human are performing, acting and creating the script of the public realm. The notion of agencies that applies to objects, infrastructure and other networks becomes the trajectory necessary to understand the mutual influences between actuators, actors and agents in the socio-technical space.

For architects and planners, this capacity of defining new actor-agent relationships to reactivate the public space expands the possibilities in designing public spaces in cities. In such a design process it will be necessary to understand what an object does, its role in the space and how the form itself will become a spatial agent, an action. The urban environment grows or changes because of the active forms within it.[19] Sociologist Manuel Castells argues, "Everything we do, from when the day begins until it is over, we do it with Internet. The connection between in-situ and virtual is established by us. There are not two different societies; there are two kinds of social activities and relations within ourselves. We are the ones that have to search the best way to arrange and adapt them."[20]

As the network offers different forms of social space, a variety of web-communities are created, proposing different models that could replace, integrate or expand traditional models of public encounters and gathering. These new models will not threaten the importance of public space but, on the contrary, foster new models of public-spatial organization through a more hybrid urbanity.

The Internet provides the tools and technology needed to claim the public that is leading towards models where the collective is empowered in the construction of the common and the shared. Participatory processes are enabled, and the Internet is the catalyst. If the construction of the public space is inherently collective, the network is then able to accelerate the process. This requires considering the parameters and potentials offered by hybrid networks when producing scenarios for cities and public spaces. Hybrid networks provide opportunities for public empowerment and for the citizen to transform public space. The currently dominating top-down urban systems will be over time increasingly confronted by local bottom-up actions. In this way the sentient city of hybrid networks can overcome homogeneity and promote a different mix between public and private activities and urban spaces.

Michael Batty and Andrew Hudson-Smith argue for the catalytic potential of bottom-up in their essay "The Liquid City": "Our understanding of how cities function is predicated on action from the bottom up. Cities are built by actions exercised by individuals on behalf of themselves or larger collectives, agencies and groups mainly configured as local actions. Global patterns emerge, best seen in how different parts of the city reflect the operation of routine decisions which combine to produce order at higher and higher scale."[21]

Mimi Zieger states that tactical urbanism uses the city as a site of experimentation, deploying pop-up parks, vacant retail reuse, or unsanctioned street furniture as ways to reprogram the urban realm. The practice traditionally takes an activist position in relationship to environmental, political, cultural and economic factors. However, as the practice is increasingly being absorbed into mainstream thinking on cities, it is critical to look closely at both the underlying assumptions and resulting effects.[22]

According to Dan Hill, the use of digital and interactive technologies should focus on transparency, open processes and open access to information, as these aim at a more human understanding of the city.[23] The changes derive from the user, from the system of networked and coordinated actions that have the potential to reprogram the software of the city.

In recent years a wide array of projects has been developed with the aim of experimenting with digital-tactical intervention. Although situating and acting locally, these can become prototypes that through the network acquire global relevance for the issues and processes implemented. Such projects propose the integration of the network within design parameters and act as bottom-up models to demonstrate the potential of information as a catalytic value added to the physical layer.

How the integration of information between citizens and systems can be organized is demonstrated by the Amphibious Architecture project (fig.5). This project was developed by David Benjamin and Soo-in Yang with Natalie Jeremijenko and sponsored by the Architectural League of New York in 2009 for the "Toward the Sentient City" exhibition. It is a floating intervention that provides an interface between life above water and underwater. Two networks of floating interactive tubes, installed at sites in the East River and the Bronx River, house a range of sensors under water and an array of lights above water (fig.6). The sensors monitor the water quality, the presence of fish and human interest in the river's ecosystem. The lights respond to the sensors and create feedback-loops between humans, fish and their shared environment.

Fig.5: Amphibious Architecture by Benjamin, Yang and Jeremijenko.

An SMS interface allows citizens to receive real-time information about the movement of fish via text-message and to contribute in displaying a collective interest in the environment. The project attempts to generate awareness of the water ecosystem as part of the city's fabric. Mapping and tracking the invisible and turning it into visible and quantifiable data encourage engagement and participation.

Another project is the "Serendipitor," developed by Mark Shepard who uses the network to enable people to explore a city in an unpredictable way, still creating an awareness of the built environment around us. Serendipitor is an alternative navigation app for the iPhone that helps people to find something by looking for something else. When the user enters an origin and a destination, the app maps a route. As the user navigates the route, suggestions appear for possible actions to take at given locations within step-by-step directions. It is designed to introduce small slippages and minor displacements within efficient routes. [24]

These and similar projects demonstrate the potential to re-program the city through the agency of the network, catalyzing processes for the generation of an awareness that might lead to long-term change through small-scale actions. Those projects act at the prototypical level; they suggest imaginable future scenarios for city growth at both the local and global scales. The tension generated by top-down versus bottom-up could be a constructive process of self-organization. The one does not exclude the other; on the contrary, both could work in a symbiotic and interdependent relationship.

"WikiPlaza," developed by hackitectura.net, aims to generate an open-space laboratory without hierarchical managed structure but instead managed by citizens (fig.7). It attempts to embody the network as a creation of a participatory public space in order to produce "ecosophic machines," that is, new technical, social and mental ecologies offering an alternative to the dominant neo-liberalism and promoting and stimulating emancipation, autonomy and spaces of the commons.[25] Through a series of equipment and tool-kits, WikiPlaza has been replicated in different cities, testing global replication through open source, participatory, self-organized, and self-managed processes. It embodies the materialization of the encounter with the Internet. The project manifests in its strategies a space that is in continuous transformation. Hackitectura.net states: "The public plaza of the future could be a WikiPlaza."[26] Projects such as WikiPlaza are able to localize the network, operating in a specific context, responding to cultural, civic and economic needs of a specific community and promoting and organizing localism." After several decades of pushing globalization-orientated values, the public tends toward processes of localization, focusing on interactions that occur at local scale. This movement between integration, centralization, globalization and

Fig.6: Amphibious Architecture by Benjamin, Yang and Jeremijenko.

Fig.7: WikiPlaza, Hackitectura.net.

regionalism including the challenge of local cultural identity acts as a pendulum in the urban decision making process.

Many similar projects have emerged from citizen governance. The "île Sans Fil"[27] project in Montreal provides free wireless access throughout the city. The short period of free Wi-Fi access has catalyzed a series of artistic and community projects that emerged from the presence of the network. The "île Sans Fil" an example out of many, implemented in an urban context, is steadily increasing the already high rate of public participation.

Very often projects initiated by networks develop an innovative hardware (spatial armature) that embodies the invisible software (programming). This family of interventions fosters participation and dynamic collaboration and sometime suggests forms of urban management. Bottom-up strategies empower citizens and are opening up the collective awareness established by shared and coordinated actions. Forms of organization take the form of the network itself.

Those types of interventions carry inherently the ability to be agents, actors and performers. The object or form is not as relevant as the potential of its impact. In this framework, architects, urban planners and designers have the potential to shift their focus. While still being concerned with geometry, materials and tectonics, they can move beyond the conception of form as object; rather they are partially the authors of form as an intense set of actions and relations deployed in space.

This shift is reframing the role and processes of spatial practitioners; it will lead to another mode of conceiving design and its methodologies. Designers will have to embody increasingly the role of facilitators of actions that they set in place and allow to unfold through multi-layered strategies. Thus, the notion of authorship changes to the coordination of networked intelligences facilitated through the project, from the "signature" that belongs to the author. It is the mutual interchange of designing spatial organizations, systems and relationships that leads to the connection between form and action.

Actions require context to be deployed, and spaces require programming to be activated. Networks are active agents as they embed actions in their protocols. The production of urban space is directed towards a territory of synthesis between object and action. Cities are assemblages of complex conditions and systems where the physical layer is continuously re-written by the constructive tension between local and global conditions. The challenge for spatial practitioners is to rethink new models of public space that act simultaneously as local - global, physical - digital, therefore hybrids. The localization of the network is the next challenge.

1. A. Burke, T. Tierney (2007) *Network Practices*. Princeton Architectural Press.

2. Castells, M. (2000) *The Information Age: Economy, Society, and Culture*. Malden, MA: Blackwell.

3. Van Dijk, J. (2012) *The network society*. London: Thousand Oaks Calif. SAGE, 3rd Ed.

4. Martin, J. (1978) *The wired Society*. Englewood Cliffs, N.J: Prentice Hall.

5. Mitchell, W. (1996) *Space, Place and Infobahn: City of Bits*. Cambridge: The MIT Press.

6. Batty, M., Hudson-Smith, A. (2012) 'The Liquid City' in *Systemic Architecture*. London: Routledge.

7. Stickells, L. (2006) ' Flow Urbanism' in *Heterotopia and the city: pubic space in a post-civil society*, London: Routledge.

8. Virilio, P. (1997) *Architecture in the Age of its virtual disappearance*. Cambridge: The MIT Press.

9. Koolhass, R. (1995) *'The Generic City'* in *S,M, L, XL*. New York: The Monacelli Press.

10. Deleuze, J. Guattari, F. (2000) *A Thousand Plateau*, University of Minnesota Press.

11. Katz, B. (2002) *Smart growth: The future of the American metropolis?* CASE paper 58. London: Centre for Analysis of Social Exclusion, London School of Economics.

12. Fuller, R. B. (1938). *Nine Chains to the moon: An Adventure Story of Thought* (First ed.). Philadelphia: Lippincott Publisher.

13. McLuhan,M. (1962) *The Gutenberg Galaxy: The Making of Typographic Man*. University of Toronto Press.

14. Weiner, M. (1991)*The Computer for the 21st Century*, Scientific American: 94-100.

15. Macmanus, R. (2009) *Pachube: Building a Platform for Internet-Enabled Environments*. Available at: http://readwrite.com/2009/05/03/pachube_internet-enabled_environments#awesm=~ooDfnun2X52u2d (Accessed: 15 June 2018).

16. Merriam-Webster, *Sentient*. Available at: http://www.merriam-webster.com/dictionary/sentient (Accessed: 15 June 2018).

17. Ratti,C. (2009) *Digital Cities:Sense-able urban design*. Available at: http://www.wired.co.uk/magazine/archive/2009/11/features/digital-cities-sense-able-urban-design (Accessed: 15 June 2018).

18. Latour, B. (2005) *"Reassembling the Social: An introduction to Actor- Network Theory"*. Oxford University Press.

19. Easterling, K. (2011) 'The Action is the Form' in *Sentient City*. MIT Press.

20. Castells, M. (2005) *The Network Society: A Cross-Cultural Perspective*. Edward Elgar Pub.

21. Batty, M., Hudson-Smith, A. (2012) 'The Liquid City' in *Systemic Architecture*, Routledge, London.

22. Zieger, M. (2011) *City Sessions: Four Questions on Tactics, Urbanism, and Practice*. Available at: http://city-sessions.tumblr.com (Accessed: 15 June 2018).

23. Baraona, E., Gonzalez, P. (2011) *About Tactical Urbanism*. Available at: http://www.dreamhamar.org/2011/09/about-tactical-urbanism/ (Accessed: 15 June 2018).

24. *Serendiptor*. Available at: http://serendipitor.net/site/?page_id (Accessed: 15 June 2018).

25. *Wikiplaza*. Available at: http://wikiplaza.org/index.php/Página_principal (Accessed: 15 June 2018).

26. *Hackitectura*. Available at: http://mcs.hackitectura.net/tiki-index.php?page=Wikiplaza+Paris (Accessed: 15 June 2018).

27. Île sans fil deviant. Available at: http://www.ilesansfil.org (Accessed: 15 June 2018).

CASE
STUDIES

ECO-BOULEVARD, AIR TREE
Ecosistema Urbano

COSMO
Andrés Jaque
Office for Political Innovation

DATAGROVE
Future Cities Lab

WENDY
HWKN

iLOUNGE
Marcella Del Signore (X-Topia)
+ Mona El Khafif (ScaleShift)

LUMEN
Jenny Sabin Studio

LIVING LIGHT
The Living

iWEB
ONL / Kas Oosterhuis

BUBBLES
FoxLin

SKIN
Gernot Riether + Damien Valero

CLOUD
Single Speed Design

BLUR
Diller Scofidio + Renfro

SERPENTINE GALLERY PAVILION
OMA - Office for Metropolitan Architecture

DIGITAL WATER PAVILION
Carlo Ratti Associati

ECO-BOULEVARD, AIR TREE

Ecosistema Urbano

The Air Tree is part of the Eco-boulevard, an urban space in Spain designed to not only generate social interaction but also to address environmental issues. The project is located in the center of Vallecas, a suburban housing development south of Madrid. The boulevard is a wide pedestrian open space that stretches over five housing blocks and is furnished by three large cylindrical structures called Air Trees which use greenhouse technology to cool the public space underneath and around them by 8-10° C during summer months.

The vertically stacked trees provide shade and are part of larger eco-technological systems. Solar cells provide not only the energy for lighting, sensors, and other equipment used to power the urban space, but they also produce a surplus of energy to support the housing surrounding the Eco-boulevard.

Each of the tree structures shares the same armature that is programmed and equipped with different technology such as fans, nebulizers, and elements that activate the space around them. One of them has swings hanging from the structure to attract children, another is equipped with features that invite people to relax, hang out, or meet up within the dynamic public space. Since Air Tree can be customized to respond to different urban conditions, the project can be inserted into any urban space that lacks amenities required for an active public space. As such the project is a prototype to catalyze, activate and improve urban space through place-making strategies.

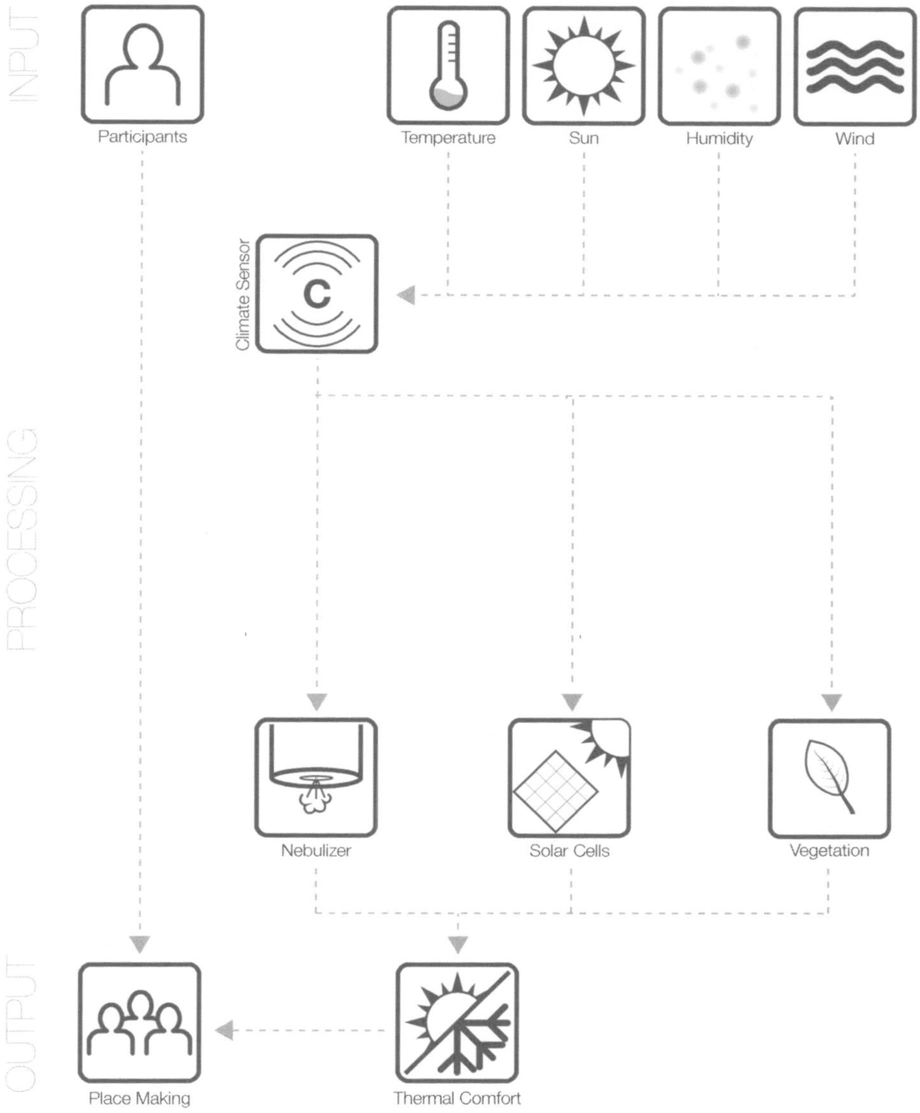

INPUT

Participants

Temperature

Sun

Humidity

Wind

PROCESSING

Climate Sensor

C

Nebulizer

Solar Cells

Vegetation

OUTPUT

Place Making

Thermal Comfort

COSMO

Andrés Jaque
Office for Political Innovation

Cosmo is a temporary installation for MOMA PS1 that reveals and makes visible a water filtration system, typically hidden in urban environments. The biochemical machine is designed to filter and purify 3,000 gallons of water. The main idea for the project is to trigger awareness of water pollution and shortage while at the same time define an outdoor space for leisure and fun.

The project is made from pipes, tanks, hoses, irrigation structures, and plants that are held in place by a minimal lightweight structural system. It uses polluted water discharged from the East River that is close to the site. The water is first poured into four connected cylindrical tanks that sit on a wheeled platform and create the base of the structure. Within the tanks, different ecosystems eliminate suspended particles and nitrates. From there the water travels through a coil of transparent pipes where it is exposed to UV light.

A series of cells filled with algae is used to reduce the water's level of hydrogen and phosphorus. The water then travels through a series of plants and filters and travels back down the structure in three waterfalls used to increase the level of dissolved oxygen. The moment the contaminated water is purified through biochemical systems, its stretched-out plastic mesh glows as a signal of the purification process. In addition, the project reduces the temperature around it by six degrees and provides shade for the outdoor party space of PS1.

People can also gain live information from the project via a webpage and, through embedded sensors, collect quantifiable data to generate community awareness around water-related issues. The intervention is an open prototype that through community appropriation can be replicated at different scales and in different contexts.

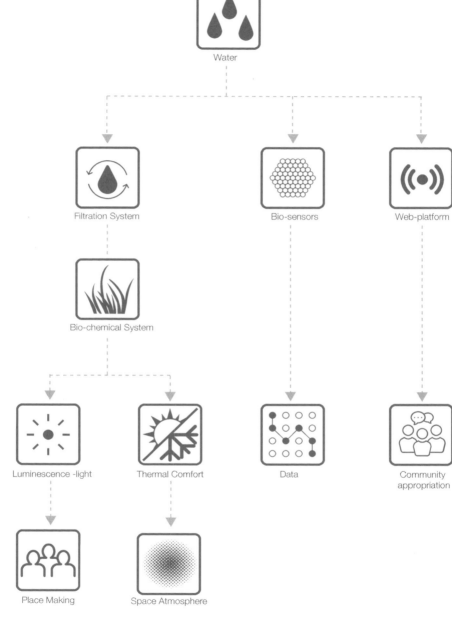

INPUT

Water

PROCESSING

Filtration System

Bio-sensors

Web-platform

Bio-chemical System

OUTPUT

Luminescence -light

Thermal Comfort

Data

Community appropriation

Place Making

Space Atmosphere

DATAGROVE

Future Cities Lab

Datagrove is located in a small courtyard in the historic California Theater at South First Street in San José. Connecting to social media the project becomes a source of information by sensing and responding to people in its immediate proximity. It functions as a social media "whispering wall" that harnesses data that is normally nested and hidden in smart phones, and amplifies this discourse into the public realm.

Datagrove is produced using advanced digital fabrication techniques and integrates a range of custom electronics including sensors, text to speech modules, LEDS and LCDS. The installation is constructed from a stainless steel and acrylic woven lattice that encapsulates five vaccum-formed LCD screens. LEDs are embedded in the acrylic tubing and the orbs, which are triggered relative to the proximity of viewers.

Sensors are used to detect the proximity of people. LCD displays that are woven into the structure start to light up as the visitor gets closer. As the public is drawn into the spatial experience of the installation, text to speech modules integrated into the structure activate whispering messages from popular Twitter feeds to the visitors.

Datagrove articulates the local and global debate as atmospheric phenomena of varying intensities of light and sound in a public space. The project illustrates how technology, social media and the Internet can be used to affect how citizens live, work, communicate and play while fostering local and global critical discussion and debate.

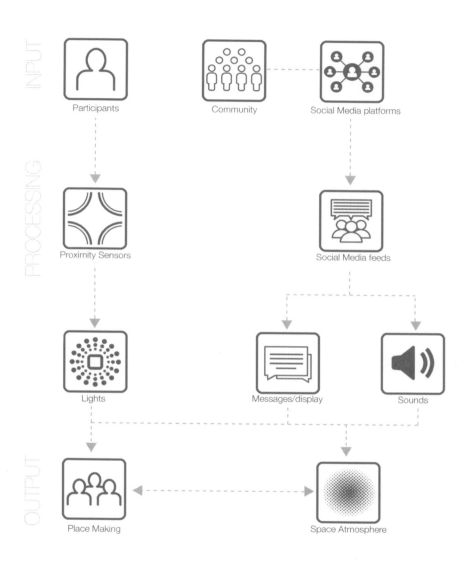

INPUT

Participants

Community

Social Media platforms

PROCESSING

Proximity Sensors

Social Media feeds

Lights

Messages/display

Sounds

OUTPUT

Place Making

Space Atmosphere

WENDY

HWKN

Wendy is a temporary intervention designed by HWKN for MoMA PS1. The project engages visitors in a new type of social environment while creating a micro-climate and filtering air pollution. Located at the edge of PS1's courtyard, Wendy creates a strong visual identity from outside while defining an event space within the courtyard of PS1.

A construction scaffolding defines the 70' x 70' x 45' space for Wendy and serves as an armature for the bold and spiky bright blue nylon fabric soft skin. The fabric, coated with a nanoparticle spray, neutralizes pollutants in the air using a photo catalytic reaction. The array of iconic spikes maximizes the surface area to increase Wendy's ability to clean the air. The pointy arms also provide shade and blasts of cooling mist and energizing music to define social zones throughout the courtyard.

Wendy's digital climate sensors trigger fans and misting devices to control the temperature of the event space. Digital responsive systems are activating social interactions within the public space. Motion sensors trigger music and water cannons to act as a social catalyst within the space.

A staircase under the spikes leads to the interior of the project allowing a person to retreat from the busy event space of PS1 and revealing most of the technology including large fans used to cool the environment. HWKN's project is a prototype that could occupy any urban environment and can be programmed to generate new complex interactions between a public space, its climatic conditions, and its occupants.

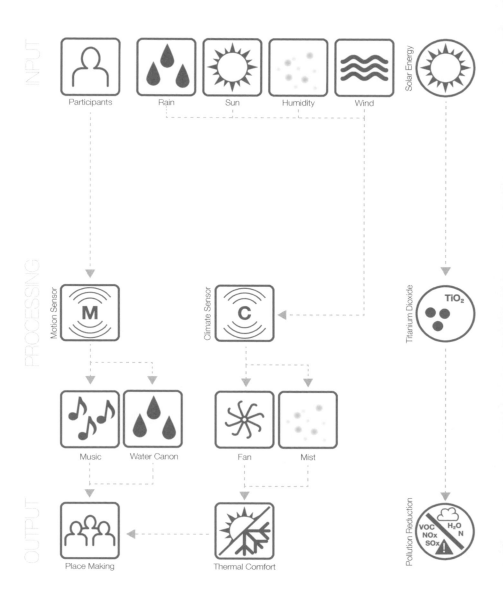

INPUT

Participants Rain Sun Humidity Wind Solar Energy

PROCESSING

Motion Sensor M Climate Sensor C Titanium Dioxide TiO₂

Music Water Canon Fan Mist

OUTPUT

Place Making Thermal Comfort Pollution Reduction

VOC NOx SOx H₂O N

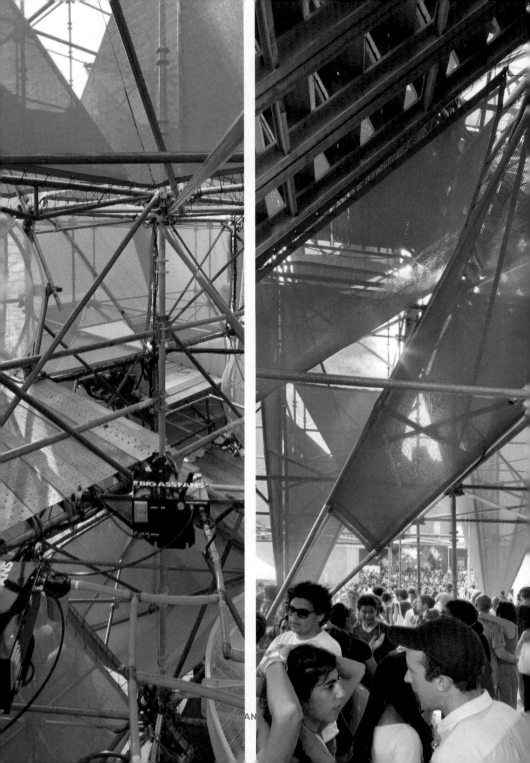

iLOUNGE

Marcella Del Signore (X-Topia)
+ Mona El Khafif (ScaleShift)

iLOUNGE is an interactive urban space, developed for Northern Spark in Minneapolis and for the ZERO1 Biennial in San José. It is an instant social stage that creates a temporary community for a minute, an hour, or an evening. iLOUNGE stimulates citizens to look, listen, exchange, reflect, and to relax. The dynamic and adaptive topography embraces and stimulates exchange by creating new connections with the existing city while engaging live feed responsive processes to augment or alter social interactions in public space.

Live feed video cameras create a media echo of the spatial production. Cameras and screens are embedded in the truncated pyramids creating a network of inputs (cameras) and outputs (screens). The cameras capture in real time the temporary space and send the information to an adjacent screen.

The nature of the space changes not only because of the tactile contact with the space, but also because of the overall game of "seeing "and "being seen" triggered by the camera-screen relations. Also macro-scale digitally mediated interactions are provided. An infrared camera is placed outside iLOUNGE to capture in real time the space in its entirety. The media footage feeds into a beamer station that projects the production of interim social space onto the surrounding firewalls. Visitors, aware of being part of the live recording, become actors participating in a process of being literally "on stage."

iLOUNGE encourages the creation of temporary communities that coexist in the physical and digital space. Analog and digitally mediated narratives that trigger interaction are embedded in the spatial construct to provide a new context for social and physical exchange.

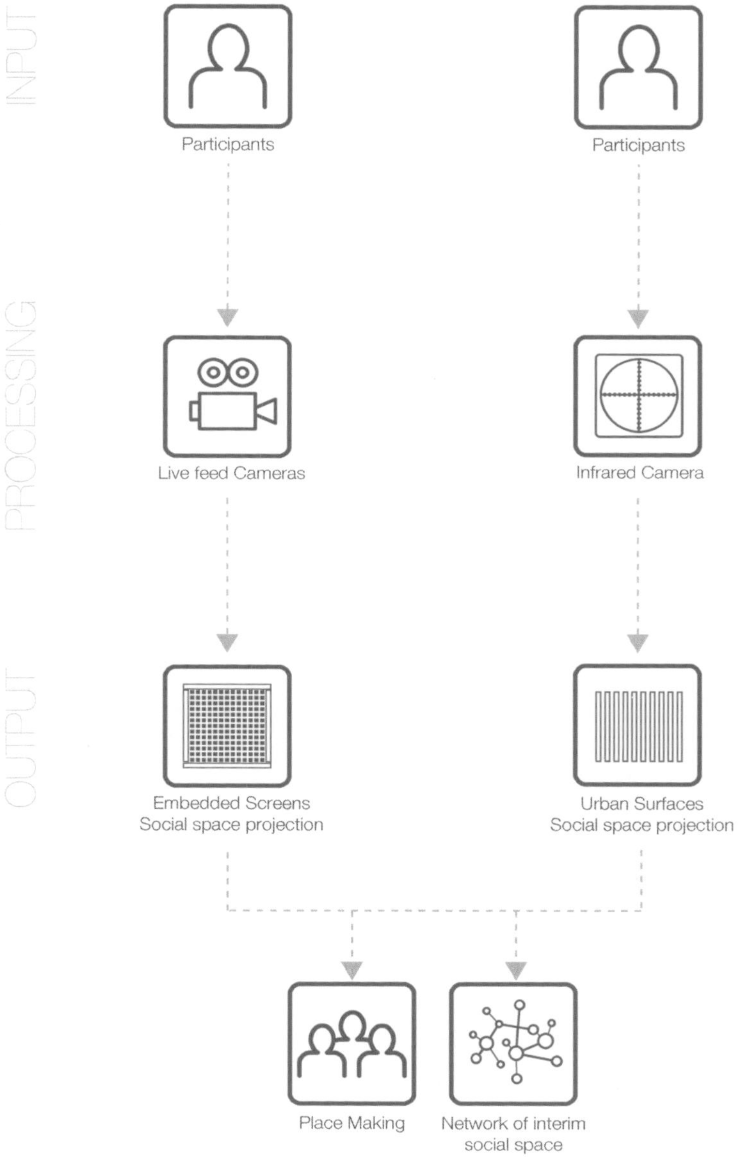
INPUT
PROCESSING
OUTPUT
Participants
Participants
Live feed Cameras
Infrared Camera
Embedded Screens
Social space projection
Urban Surfaces
Social space projection
Place Making
Network of interim
social space

take off your shoes and become an actor

off your shoes and become an actor

you are on camera... smile! :]

take off your shoes and become

you ar... smile! :]

LUMEN

Jenny Sabin Studio

Lumen is a socially and environmentally responsive structure that adapts to the densities of bodies, heat, and sunlight. The dynamic structure was developed to change the spatial qualities of the PS1 MOMA courtyard while providing an immersive experience for the visitors. By night, Lumen is bathing visitors in a responsive glow of photo-luminescence; by day, Lumen offers succor from the summer heat, immersing participants in delicious ground clouds of cooling mist.

Over a million yards of thread were robotically knitted into a network of hanging tubular cells that define a network of public spaces. Each cell varies in length and proportion. The cells are horizontally aggregated to form a spatial landscape of stalactites that occupy the entire public courtyard of PS1. The individual cellular components are elastic and can adapt to the geometry of the neighboring cells forming a light and soft pattern that is held in tension by the heavy rigid concrete walls that form the courtyard.

Movement, temperature and sunlight activate the fabric stalactites. The cells are made from solar active fibers that change color when exposed to sunlight. Spaces dramatically change in characteristics, from a responsive glow of photo luminescence to a misty cloud. Families of stools populate the ground plain. Their locations and sizes respond to the tubular structures of the canopy. Sensors imbedded in the stools are used to detect the motion of visitors and inform a misting system that cools down the environment locally and provides visitors with relaxing spaces during the hot summer in New York.

Through direct references to the flexibility and sensitivity of the human body, Lumen integrates adaptive materials and architecture where code, pattern, human interaction, environment, geometry and matter operate together as a conceptual design space.

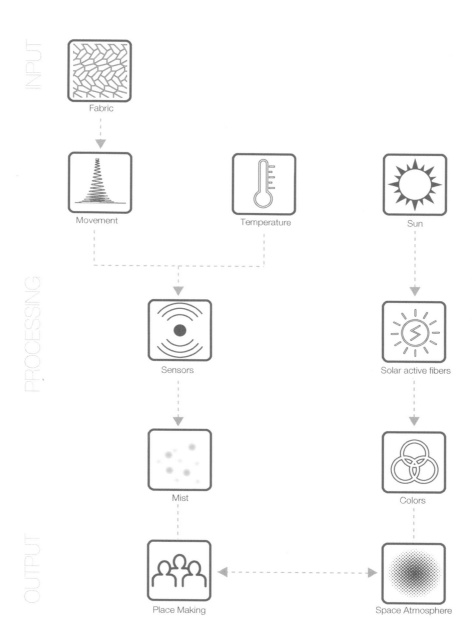

INPUT

Fabric

Movement

Temperature

Sun

PROCESSING

Sensors

Solar active fibers

Mist

Colors

OUTPUT

Place Making

Space Atmosphere

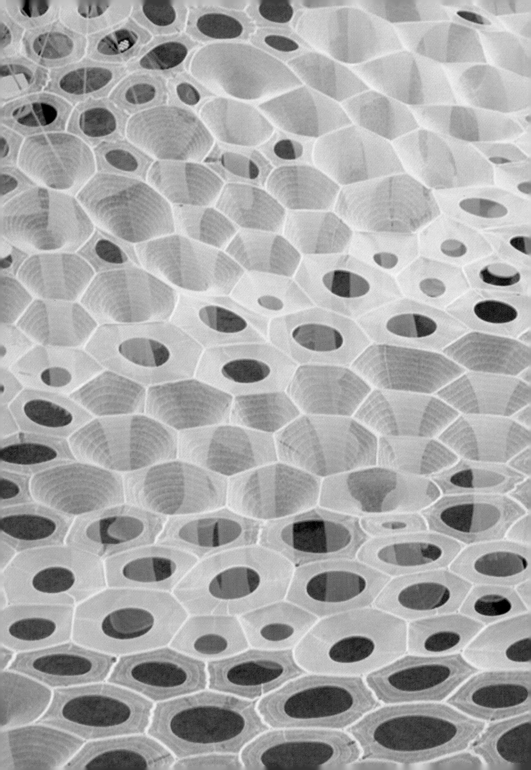

LIVING LIGHT

The Living

Living Light is an outdoor canopy in Peace Park, across from the World Cup Stadium in Seoul, South Korea. Using the high visibility of the site, the responsive canopy informs the visitors about the city's air quality.

The glass-panel canopy is supported by a tree-like steel structure. Each panel represents one of 27 Seoul neighborhoods and is connected to an air monitoring station in that neighborhood. The Korean Ministry of Environment operates the air monitoring stations which send real-time air quality information to the canopy. The individual panels are illuminated differently based on the improvement in air quality of individual neighborhoods as well as best and worst air quality of neighborhoods. The panels blink once the system sends requested reports to citizens via text message.

The project was designed and conceptualized as a prototype for a building facade that would visualize information about the city's environment. As many buildings in Seoul are already lit up at night by large billboards used for advertisement, the Living Light project suggests a facade that will make citizens more aware of pollution, consequences of consumerism, and environmental issues. Through the pavilion, citizens participate in a public discussion about the future of the environment. The project suggests that the facade strategies of buildings could also inform citizens about issues that matter in real time.

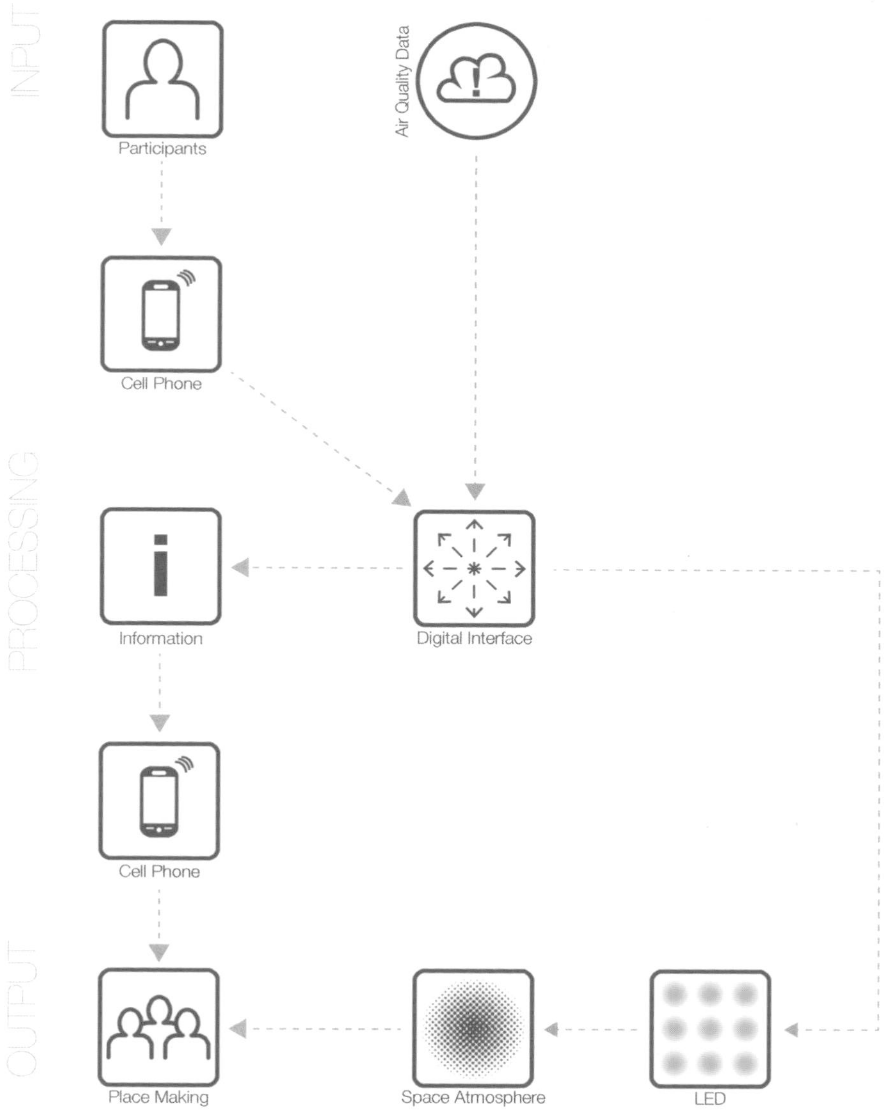

INPUT

Air Quality Data

Participants

Cell Phone

PROCESSING

Information

Digital Interface

Cell Phone

OUTPUT

Place Making

Space Atmosphere

LED

iWEB

ONL / Kas Oosterhuis

The iWEB pavilion is an armature for collaborations necessary to tackle complex issues such as climate change. Assembled by robots, the iconic sculptural object is a laboratory for the public, a real-time digital design platform that allows unlimited mass collaboration.

Assembled by two industrial ABB irb 6400 robots, the pavilion is able to erect itself in any public space. The complex monolithic object's surface is tessellated into triangular, flat, and folded steel plates. Bright light penetrating through gaps between the panels suggests its living interior. As only a single group of panels fold outwards, the pavilion reveals its architectural quality as a space that can be entered. The interior is a single 250-square-meter space equipped with information technology to facilitate communication such as speech recognition or hand movement.

Once inside, every visitor is considered an equal "player." The visitor has the option to trigger a pressure pad that puts him forward as a player within an abstract design environment. Then the visitor is asked to participate in tackling relevant environmental problems and the design of sustainable solutions. Information is exchanged through five large screens and a multi-channel audio system, which allows for collaborations in real time. The multi-player design environment is a vehicle for trans-disciplinary research, exchange of ideas, debates, education, and design.

The project is activating the public by allowing an easy transfer of knowledge. The project enhances the sensitivity of players to each other's points of view. It allows them to design in real time and collect solutions. As an interactive architecture, the pavilion can also be used for public lectures, walk-in presentations, discussions, and parties and can serve as an icon for communal initiatives in any type of public space.

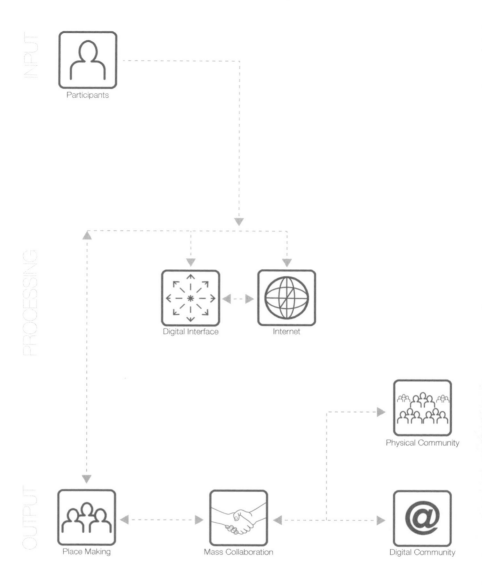

INPUT

Participants

PROCESSING

Digital Interface

Internet

Physical Community

OUTPUT

Place Making

Mass Collaboration

Digital Community

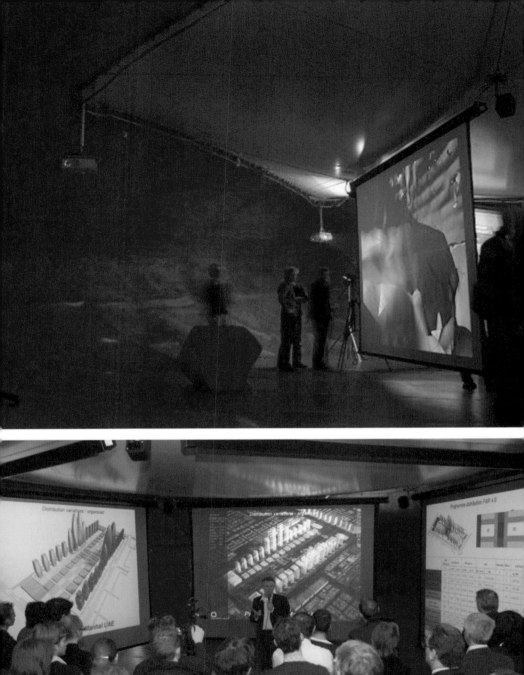

BUBBLES

FoxLin

Bubbles is an adaptive spatial pneumatic environment that occupies the pocket park of the Materials & Applications Gallery in Silver Lake, Los Angeles. The park, a former parking lot, is public and open to the street. The project entirely consumes the park and invites the public to explore a novel interactive spatial condition.

The installation consists of 16 large pneumatic translucent rip-stop nylon airbags, or bubbles, each eight feet in diameter. The airbags hang by clear ducting from a support structure. At the center of each bubble is a capsule that contains a micro-florescent lighting element to make the bubble glow and a sensor used to trigger the fan that can inflate or deflate the bubble.

When the bubbles are touched, sensors initiate an exchange of air between the bubbles. When fully inflated, the bubbles fill the entire space. As visitors come closer, the bubbles deflate and create spaces that can be occupied by individuals or larger groups. Each bubble responds to visitors and to neighboring bubbles. One bubble responding to an occupant can trigger a chain reaction causing activity in the entire field of bubbles. When the site is unoccupied, the bubbles slowly inflate back to a state of equilibrium, a stand-by state that fills the entire site until another interaction takes place.

The installation promotes a public space that is adaptive rather than a static environment. The intent of the project is to create a fully immersive architectural environment that could spatially respond to changing social conditions. The bubble-to-bubble and bubble-to-human interactions create a complex environment in which emerging spaces are difficult to predict. As such the project encourages a dynamic relationship between citizens and urban space.

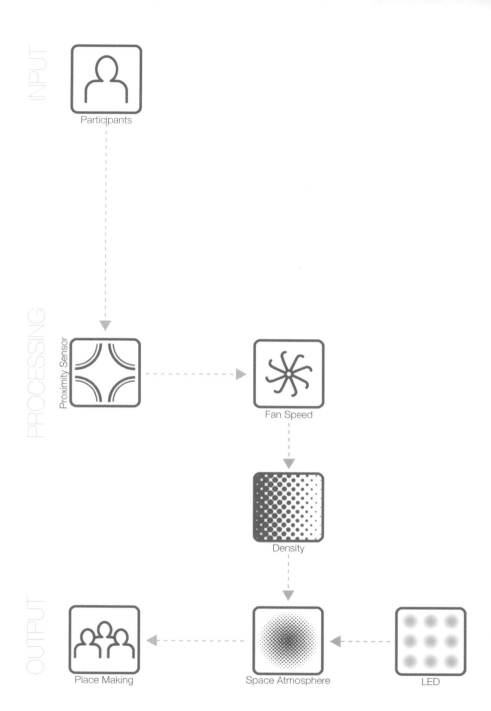

INPUT

Participants

PROCESSING

Proximity Sensor

Fan Speed

Density

OUTPUT

Place Making

Space Atmosphere

LED

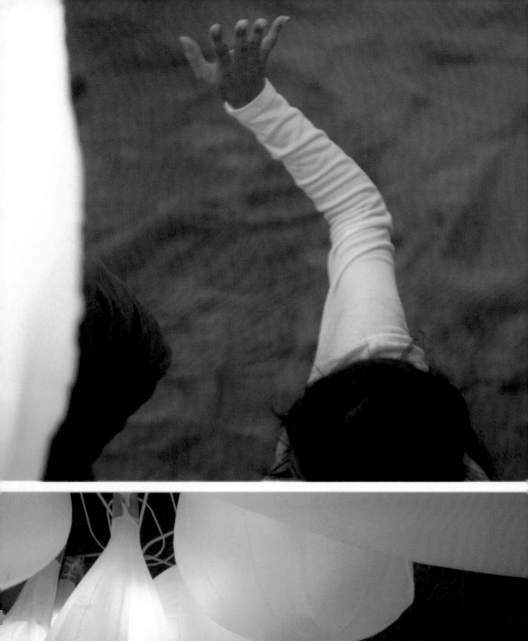

SKIN

Gernot Riether + Damien Valero

SKIN was built for the Traverse Video film festival in Toulouse, France. The project uses video and sound to provoke the relationship between the human body and the physical environment of public space. The multi-layered cellular structure of SKIN is informed by data taken from the human body. Engaging with the project the spectator is able to reconstruct an artificial environment from transplants of videos of human skin and sound fragments collected from medical devices that capture sounds within the human body.

The epidermis of the human skin is translated into a series of layers that form a self-supporting structure. The layers are constructed from flattened elastic polymer cells that, depending on the curvature of the overall form, change in size and thickness. Similar to the human skin the envelope serves as a regulatory device that interacts with the exterior and interior context.

The material for the video is derived from a scan of the skin of a female and male body. This footage is then organized into transplants that are scaled to match the size of the individual cells of the project. Some cells are equipped with tactile sensors that allow the spectator to edit the visual effect of the space. The movement of the spectator is translated into acoustic effects that were originally captured from different locations within the human body through media devices and edited for the SKIN project by sound artist Jérôme Pougnant.

Through proximity sensors the visitors are able to reassemble the sound fragments into new acoustic environments. The spectators become actors by involving their own bodies in a relationship with the sensory object. It suggests public spaces that are defined by surfaces that have depth, are multi-layered and are highly interactive.

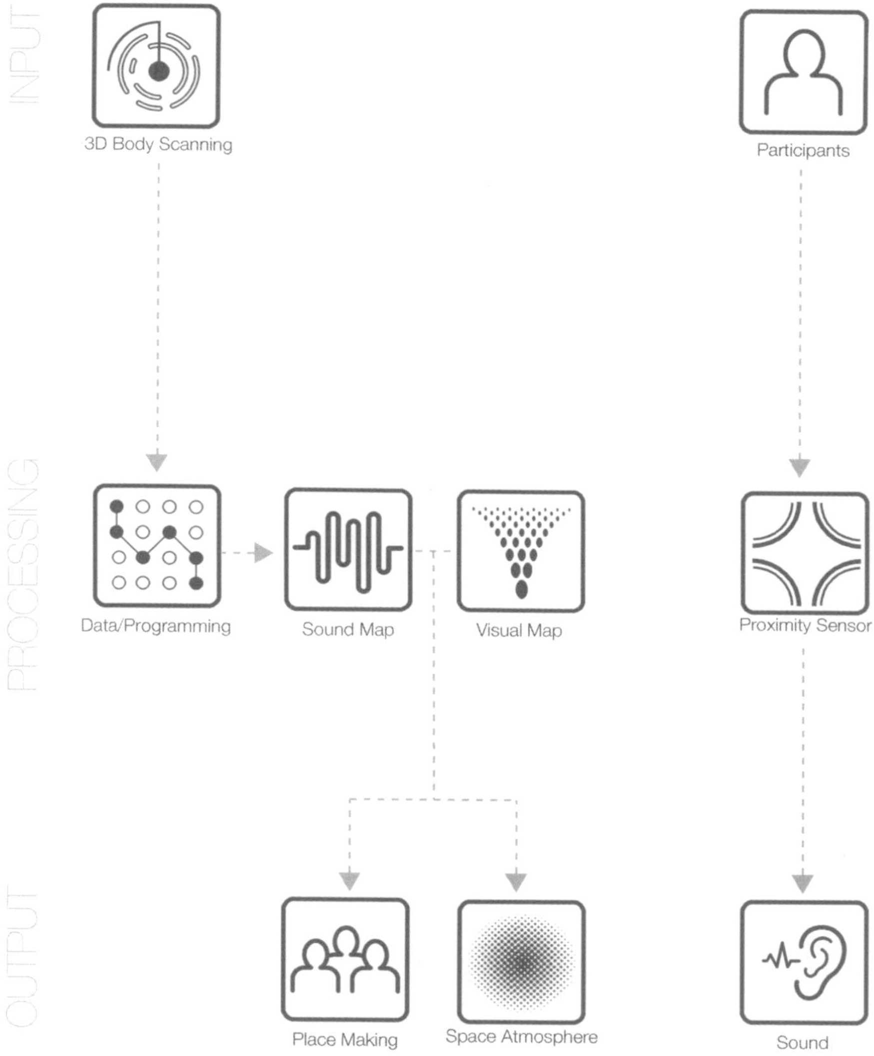

INPUT
3D Body Scanning
Participants
PROCESSING
Data/Programming
Sound Map
Visual Map
Proximity Sensor
OUTPUT
Place Making
Space Atmosphere
Sound

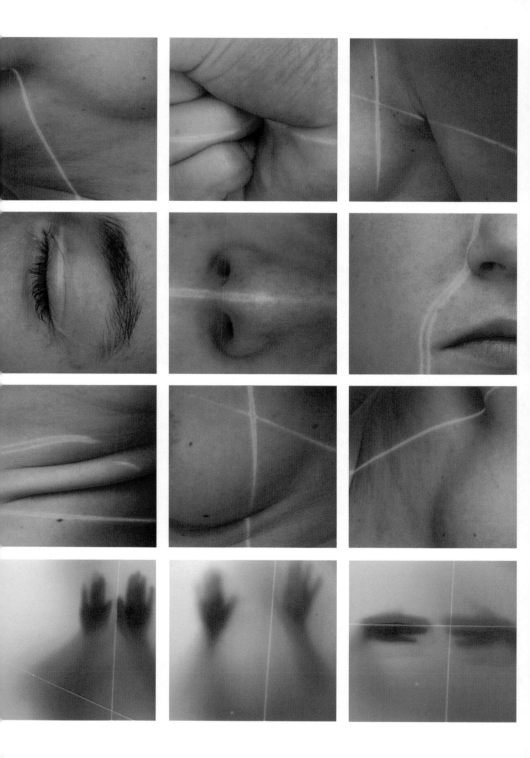

CLOUD

Single Speed Design

Cloud is a permanent interactive installation at Heyri Art Valley, an artists' village in Paju, South Korea. The project consists of a series of canopies that engage pedestrians in a playful interaction with data collected from changing weather conditions. The main goal of the project is to enhance the experience of a public pathway between a street and a riverfront and to draw attention to the artists' village.

A linear structure of diagonal aluminum columns supports a series of canopies in a cloud-like formation. Each of the clouds is made from an array of lights and speakers that respond to local weather conditions and movement of pedestrians. Aluminum channels house all the electronics and acrylic rods used to amplify the LED lights. The canopies come to life when people approach. An ambient whispering invites people to come closer. Once sensors detect an individual or a group of people, the information of their movement is sampled with information collected from temperature, wind speed, and humidity.

This information is often translated into surprising effects. Low temperatures, for example, turn LED lights into an orange field; warmer weather changes into patches of blue lights. The goal of the project is not only to visually communicate information about weather conditions but also to use this information to enhance the quality of public space. The resulting sound and light patterns link the public space, natural environment and visitors to produce continuously changing atmospheres, a playful dynamic interaction that not only draws attention but also creates a new identity for the artists' village.

URBAN MACHINES: Public Space in a Digital Culture

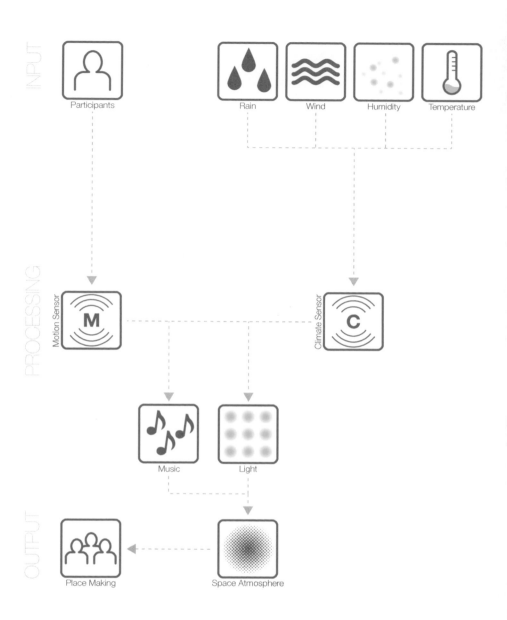

BLUR

Diller Scofidio + Renfro

Blur is the centerpiece pavilion of the sixth Swiss Expo. The temporary structure is a man made cloud that hovers above Lake Neuchâtel. The cloud suspends the visitors in white mist that blurs the environment and temporarily disconnects them from it. The project is a critique on media's high-resolution images and sound.

A long bridge leads from the bank of the lake to a 200' x 300' platform located within a lightweight structure that is supported by four columns. Water is pumped from the lake, filtered, and shot as fine mist through 35,000 high-pressure nozzles that occupy the structure. The water droplets produced by the nozzles are small enough to remain suspended in the air creating a dense fog that erases any visual reference of the environment. The nozzles also produce a mesmerizing sound that heightens this effect by blocking out any acoustic references as well.

Information about temperature, humidity, wind speed, and direction is used to regulate the water pressure of the nozzles to adjust the strength of the spray according to different climatic conditions. As the cloud responds to continuously changing weather conditions, its overall form changes, and its volume expands or contracts.

Prior to entering the cloud, visitors are handed raincoats called "brain-coats." The raincoat can change in color to show antipathy or affinity to other visitors on the basis of a questionnaire that visitors fill out prior to entering the cloud.

The project is using information technology to question the materiality of architecture and the boundary and definition of public space allowing visitors to detach from the typical media-dominated environment and experiment with new forms of social interactions.

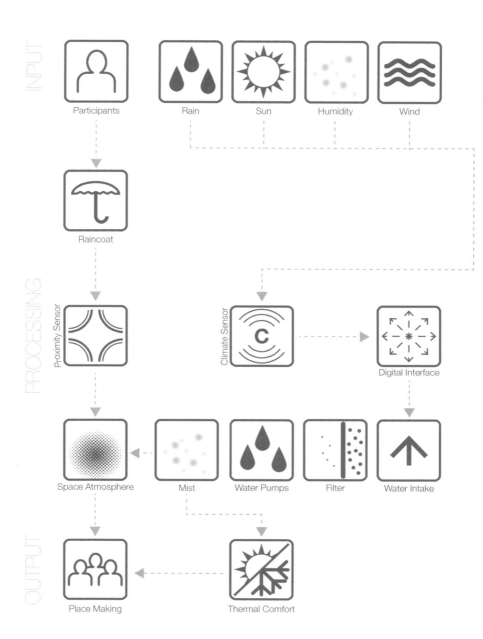

INPUT

Participants · Rain · Sun · Humidity · Wind

Raincoat

PROCESSING

Proximity Sensor · Climate Sensor · Digital Interface

Space Atmosphere · Mist · Water Pumps · Filter · Water Intake

OUTPUT

Place Making · Thermal Comfort

SERPENTINE GALLERY PAVILION

OMA - Office for Metropolitan Architecture

OMA's pavilion is a temporary structure located next to the Serpentine Gallery in Kensington Gardens, London. It is a public space that uses information technology to facilitate the inclusion of individual visitors in a communal dialogue and shared experiences. The pavilion, which can expand and contract, is conceptualized as narrative journalism, a content machine defined by events and activities.

Located in axial relation to the existing gallery building, the structure's circular floor plan mirrors the gallery's central space. Its roof is a giant helium-filled, semi-translucent balloon made from polyester. The balloon contains 6,000 cubic meters of pressurized helium. Since the gas contracts and expands under different temperatures, the pavilion's roof is almost flat at cold temperatures and expands during warmer days. Ten cables are loosened and tightened accordingly using electrical winches to anchor the balloon to the ground and provide control during windy conditions. The balloon is illuminated at night.

The room below the balloon is a public space that functions as a café and forum. Projectors and screens are used to connect the visitors to daily televised and recorded public programs, live talks, and film screenings including 24-hour interview marathons with leading politicians, architects, philosophers, writers, artists, film-makers, and economists exposing the hidden and invisible layers of London. The space is generated with two layers of translucent polycarbonate panels: one exterior layer forming a circle, and one interior forming a square. The spatial conditions both reflect and distort the way those inside events appear from the outside. The balloon roof and the occupied space below are separated by wallpaper designed by artist Thomas Demand.

The pavilion is a prototype for a public space defined by events and activities. The project allows the public to experience a built environment that is dynamicly integrated to events around the city.

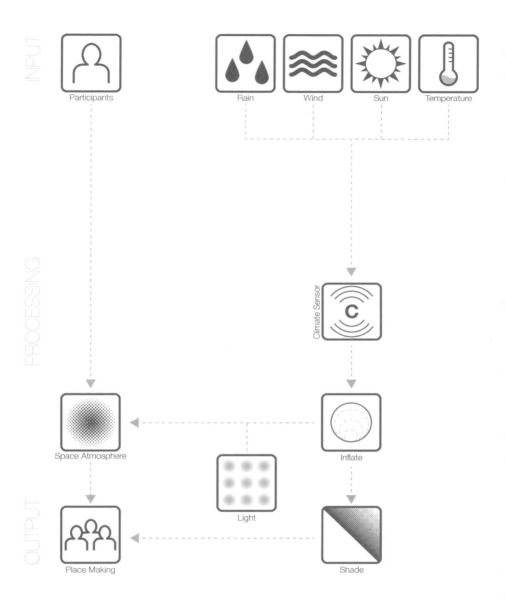

INPUT

Participants

Rain Wind Sun Temperature

PROCESSING

Climate Sensor

OUTPUT

Space Atmosphere

Inflate

Light

Place Making

Shade

DIGITAL WATER PAVILION

Carlo Ratti Associati

The Digital Water Pavilion was built for the World's Fair in Zaragoza, Spain, in 2008. Formed by interactive water curtains, the pavilion defines a space that can dynamically adjust to people and weather conditions.

The pavilion contains a temporary exhibition area, a café, and a public space for events. The only solid element of the pavilion is a roof designed as a 400 mm deep pool that collects rainwater. 3,000 digitally closely spaced controlled solenoid valves positioned all around the roof produce a curtain of falling water that entirely encloses the spatial volume of the pavilion.

The opening and closing of the valves at high frequency is digitally controlled. Changing the frequency of water droplets allows the water curtain to generate patterns and apertures. Using sensors for detecting a person who approaches the space, the system creates a temporary gap in the water curtain to enable a visitor to enter the pavilion. Touching the water surface at any point propagates patterns along the walls of the pavilion. The result is a space that is interactive and reconfigurable by water walls; the internal partitions can shift depending on the number of people occupying the space.

The roof is supported by twelve hydraulic stainless steel pistons that can move the roof up and down depending on wind conditions. This effect changes the spatial proportions underneath. When the pavilion is closed the roof flattens into the ground to decrease the building's volume. A series of pumps is used to support the water's recycling system.

The digital augmented pavilion redefines boundaries in the public environment to be reconfigured at any time. This project connects responsive information technology with water, an element that has always been an important part in the history of public space.

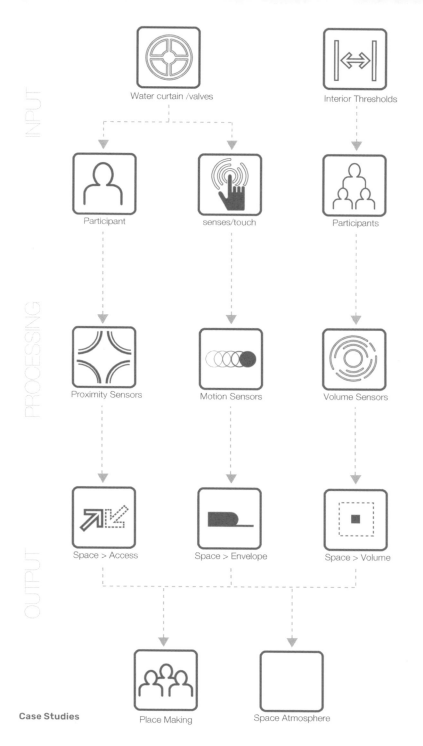

INPUT

Water curtain /valves

Interior Thresholds

Participant

senses/touch

Participants

PROCESSING

Proximity Sensors

Motion Sensors

Volume Sensors

OUTPUT

Space > Access

Space > Envelope

Space > Volume

Place Making

Space Atmosphere

CONVERSA-
TIONS

Urban Social Design, Network, Participation and Open Source City
Ecosistema Urbano

SENSEable CITY
Carlo Ratti with Matthew Claudel

Systemic Architecture
ecoLogicStudio

Cities, Data, Participation and Open Source Architecture
Usman Haque

Situated Technologies
Omar Khan

Synthetic Urban Scenarios
Areti Markopoulou – Manuel Gausa

Machines and Apparatuses as Scenario
François Roche

01

Urban Social Design, Network, Participation and Open Source City

Ecosistema Urbano

1. Sustainability and Networks are very important themes in your research and design. Some of your projects, especially those that operate more at the large city scale, are imagining a set of conditions that provide scenarios rather than a more permanent and immutable Master Plan. Why do you think the approach of Scenario Planning might have a much stronger potential to envision future urban space?

U.S.: In architecture, according to relatively recent innovative approaches such as *design thinking* and *service design* that are emerging from other disciplines, the process itself is becoming as important as the final product.

The increasing culture of process is asking architects and urban designers to become more aware of the implications of design in the different aspects of reality while allowing them to engage new inputs, incorporate feedback, and make the required adjustments on the environment they are designing, before they have finalized the project itself.

Therefore, the role of the architect is moving from only delivering built products to understanding and harnessing their capacity to cultivate projects as catalysts of ideas and mediators between stakeholders and managers.

2. What are the challenges of supporting sustainability through place making? Can you illustrate the main strategies adopted for your project titled "Bioclimatic Improvement Strategy for Public Spaces"?

U.S.: We understand empowerment as the process by which a person or a community improves its ability to act upon the environment. Empowerment is supported by the increased autonomy given by knowledge and access to the necessary tools. It can be applied to different aspects of human life such as political, technological, economic or cultural ones.

In regards to citizens, empowerment produces a fundamental change in their relationship with the environment they inhabit and with its functioning conditions. Empowered people begin to understand the dynamics, discover resources or learning tools, and develop skills that allow them to play an active and transformative role in the development and management of the environment.

"The Bioclimatic Improvement Strategy for Public Spaces" project defines a series of solutions, devices, and tools to improve the climatic conditions within the network of public spaces in Madrid. These proposed responsive environments—according to different parameters, both climatic and programmatic—reconfigure the space to achieve the best performance.

3. In some of your writings, you have talked about the notion of learning and emulating the conditions that foster the exchange and interaction that occur among users of the web to the physical urban space. How can this propose a new model for urban design?

U.S.: Communities have been gathering in defined physical places since the very beginning of human existence. With the arrival of the digital era, new types of communities have emerged. Driven by common interests and goals, communities create new places in forums, blogs, or social networks on the Internet.

Nowadays, projects that have an impact on public space and architecture have to combine the digital and the physical dimension of space. Both environments are becoming more complementary and connected, combining each other's advantages in a powerful way. Many projects are now born in a constant tension between the experience of residence and visitors, the specificity of the local context, the global environment, and the shared flow of information and knowledge.

4. How can we, through design and information technologies, improve and promote urbanity?

U.S.: Technology is a social tool, not only an instrument for problem solving. It brings new possibilities that usually change the way we behave. That is specially the case with information and communication technologies that enable us to better relate and interact with each other and with the surrounding environment. As the digital-physical divide narrows and the possibilities multiply, they are becoming an increasingly significant element in urban social life.

New social tools that help mapping, evaluating, documenting, or sharing opinions on reality are enabling participation and lowering the barriers for citizen engagement. The nearly ubiquitous presence of the Internet connection and the abundance of portable display and control devices such as laptops, tablets, and smartphones is amplifying that effect, allowing for interactions to happen almost everywhere.

On the professional side, the use of Information and Communication Technologies for "being in a network" brings another massive change. Many architects in Spain are discovering the power of social communication tools, which help generate a professional network around common interests and a mix for new initiatives. Some architects, Ecosistema Urbano among them, started following this path, progressively building a distinctive digital identity. Now the network is more established, and initiatives, knowledge, references, and reflection on operating models is even more shared. As a consequence, the instant and ubiquitous connection is accelerating the emergence of a culture of sharing versus a culture of individualized practices.

5. Why are participation and open source urban strategies such powerful tools for urban renewal?

U.S.: We live in an era of digital tools that provide new opportunities of communication and interaction. The aim is the creation of a strong community network. Citizens are empowered through the use of these tools that link different groups of population—professionals, seniors, children, adolescents, and others—in new ways.

The concept of working together in a common project improves and enhances the community. The most important goal of participatory processes is targeted towards the social, that although less visible, have the power to extend beyond the action itself.

6. Instant Urbanism is an alternative form of planning that through bottom-up processes could produce a strong impact on social processes. Some of your built projects fully embody those principles. What are the necessary parameters that make those interventions successful?

U.S.: Today technology and society have multiplied the speed of a project's transformation over time. These create challenging processes and new possibilities. In this context, the permanence of architecture is questioned. Its durability and flexibility have to be rethought. Concepts of sustainability must embody this fully for their capacity to adapt and to project future scenarios that take into account the practice of reversibility.

In addition to this, we believe that in today's connected world, a project should be the result of an open and multilayered network of creative designers, technical experts, citizens, and stakeholders that generates design from a complex set of data, needs, and inputs. Within this new context, it is necessary to explore the new role of the designer as a mediator and curator of social processes in a networked reality. The designer should try to visualize and materialize new urban models and design tools that enable the incorporation of the citizen as an active agent, avoiding the conventional paradigm of the citizen as a consumer-viewer-recipient of a finished product.

This paradigm shift, as we can see, has direct consequences in contemporary urban experiences. It is time to participate. The questions are: how, with what tools, what thought, what channels? What role can architects take? We know that as architects we have great challenges ahead, and we believe the opportunities are endless.

1. http://ecosistemaurbano.com/portfolio/bioclimatic-improvement-strategies-for-pub-lic-spaces/ (Accessed: 15 June 2018).

2. http://ecosistemaurbano.com/portfolio/air-tree/ (Accessed: 15 June 2018).

3. http://ecosistemaurbano.com/portfolio/galicia-pavilion-at-zaragoza-expo-2008/. (Accessed: 15 June 2018).

4. http://ecosistemaurbano.com/portfolio/energy-carousel/

BIOCLIMATIC IMPROVEMENT STRATEGY FOR PUBLIC SPACES

Madrid is a lively, vibrant city with lots of activities held in its public spaces. However, during the summer the city reaches high temperatures of up to 40° C, which combined with low humidity levels, makes the daytime outdoors experience uncomfortable. This situation drains activity from the public scene and forces people into the air conditioned indoors. A series of low-cost, low-tech interventions have been developed to improve and upgrade the climatic comfort and conditions of the network of public spaces in Madrid. Five strategies were identified to tackle this issue: unify (transforming fragmented and discontinuous space to provide universal access to all public spaces), concentrate (redefining urban elements for a more rational use, optimizing resources and elements), re-naturalize (incorporating natural elements into the urban regeneration process), condition (using natural and economic elements and systems to create artificial climatic comfort), activate (promoting and intensifying activity and designing open web tools capable of activating public space).[1]

Fig.1: View of bioclimatic improvement strategies for public spaces.

AIR TREE SHANGHAI

The Air Tree emerges as an experimental prototype of intervention in contemporary urban public space, capable of reactivating sites and creating the conditions to empower the use of the collective space. It is conceived as technological urban furniture, which also serves as a virtual node of connectivity. Its different technical layers enable multiple final configurations and a myriad of intermediate appearances (opaque, translucent, transparent, bright, interactive, open, etc.). Different textiles for video projections allow an unlimited combination of scenarios adaptable to citizen needs. Its appearance can be transformed over the daily cycle and through different seasons. Through sensors, it is linked in real-time with the climatic conditions of Shanghai, constantly adopting the optimal physical and energy consumption configuration to generate climatic comfort for the citizens.[2]

Fig.2: Air Tree Shanghai.

ENERGY CAROUSEL

Energy Carousel is an educative and multi-age friendly playful object. It consists of a tensegrity structure formed by ropes and textiles. The kinetic energy released by users hanging and turning on its ropes is captured by a carousel structure and stored in a battery underneath the play site. This energy is used to generate lighting effects in the evenings. The mechanism of its energy production, and lightning is as simple as a bike dynamo. The color of its lights changes based on the amount of energy generated by its users on a particular day.

The project was designed to promote education through play. Teaching children about alternative methods for generating electrical power with their own physical experiences was used to advocate a more sustainable approach to urbanism.[3]

Fig.3: Energy Carousel.

GALICIA PAVILION AT ZARAGOZA EXPO 2008

In Galicia, Spain, there are many forms of water: the sea, the ocean, rivers, springs, rain, dew. Water is present in all the elements that form the landscape of Galicia. Water is its sustenance and its wealth. It is its soul and its reason for being. The pavilion is a way for Galicia to pay homage to the element of water. There are 315 municipalities in Galicia. Each has a unique relationship with the water present in its territory. These waters tell stories about families, villages, cultures, and industries. For the project, 24 samples of water (sea, river, tap, spring, fountain, etc.) were collected from every municipality. They were then bottled in air-tight recycled PET containers and labeled to indicate origin and chemical composition. The containers were stacked to create a wall of bottled water that represents the full range of waters in Galicia as a bottled territory.[4]

Fig.4: Interior of Galicia Pavilion.

02

SENSEable CITY

URBAN MACHINES: Public Space in a Digital Culture

Carlo Ratti with Matthew Claudel

1. Cities are radically transformed by processes that engage real-time systems deployed through sensors, actuators, hand-held electronics, and devices. This allows for rethinking new strategies to conceive the built environment. How do you envision design tools to evolve in order to continue to engage these new parameters?

C.R. + M.C.: The city today is a hybrid space, caught between digital bits and physical atoms. Every dimension of daily life has some kind of virtual repercussion – whether it's a credit card swipe, a train ride, or adjusting your thermostat – which means a new playing field for designers. The key shift in the years ahead will be based on two-way communication with the built environment. As everything in the city becomes alive, in a sense, it can be monitored dynamically and controlled dynamically. Designers will have to think beyond the moment when they hand off their construction documents, and consider how people engage with what they create, as an ongoing relationship. That is, we might see architecture shift from a linear process – from designer to inhabitant – becoming something more like a feedback loop. Not so much living in buildings as living with them.

2. Cities are becoming computers through technologies that ubiquitously enter the physical realm of things. More than defining cities as "smart" which entails the removal of the subject, more often we refer to cities as "intelligent " or "sentient" where people are fundamental actors for a two way communication. How are participatory technologies that engage the individual as the agent able to change the management of cities?

C.R. + M.C.: Labels like "smart" and "intelligent" imply a very strict, specific operative framework. At the lab, we prefer to think of cities as "sense-able," as it carries a connotation of sensing and feedback. The common denominator for all our projects is that they are focused on people, rather than technology.

Connection, whether it's sharing through alternate economies or new modes of working in parallel, will provide the sea change in our cities tomorrow. Glimmers of this are appearing today, as citizens are empowered to take an active role in their environment. Humans have always worked together towards common goals, but networked tools are now allowing that to happen across space and time in an unprecedented way. Ideas can catch like wildfire and have a much broader impact. At times, the technologies (or apps or anything) that emerge from the bottom up can be disruptive to existing systems – such as Uber taxi – but ultimately, if they provide a better service for customers and an improved opportunity for providers, they are actually much 'smarter' than what we ordinarily think of as "smart city technology."

3. Within an urban environment, the phase of sensing (observation and collection of data) and the phase of actuating (acting upon a system) are two important steps to capitalize on the potential of Information and Communication Technologies as tools to catalyse urban processes. What are the challenges and opportunities to transition from Sensing to Actuating to generate a feedback loop able to produce a visible impact on cities?

C.R. + M.C.: In many ways actuating technologies – ones that have a visible and tangible impact on daily life – are difficult to imagine. What does technology have to do with my parking spot, or how warm I feel? But in fact, several of these technologies are already on the brink of entering our urban space, from autonomous driving to dynamic, user-responsive climate control. Because they are built on the same networked backbone as their sensing counterparts, each will reinforce the other. Given the right raw material – data or systems access – designers will build actuators based on the sensed data, and vice versa. The role of a city, then, is to provide the tools for citizens to contribute to a powerful, distributed innovation ecology that actuates dimensions of the immediate space around them.

4. Bottom-up processes foster empowerment and participatory forms of design. Open source strategies operate within a horizontal and distributed network. How do you envision open source initiatives connected to information technologies to change communities and city life?

C.R. + M.C.: Open source strategies stand to make a tremendous impact on both the design and operation of cities – as they already have in the fields of software (with Linux) and fabrication (with Fab Labs) for example. Open, shared, community-based design models have historically existed in building, from Italian hill towns to Gothic cathedrals to barn raising, but they have been overshadowed in modern times by the rise of the architect-as-artist. Today that is changing, thanks to networked sharing technologies.

On a platform of networked digital tools, communities might emerge, composed of members from around the world, to solve specific design challenges. Entities that need design – from disadvantaged individuals to multinational corporations – will have a means to tap the power of the crowd, and talented people will offer their expertise in response, motivated by a more flexible system of authorship, citation, and rights. Nevertheless, bottom-up processes won't just spontaneously happen, so the role of the designer will shift from a narrow design focus to more broadly encompass starting a project, nurturing its collaborative development, and synthesizing the final result – what could be thought of as a choral designer.

5. The modes of production of the public realm are continually redefined by the layering of data within the hardware of cities. As a consequence, not only urban forms will change but also modes of experiencing the public realm. Do you envision the emergence of a new typology of public space?

C.R. + M.C.: Humans will always need the same basic spatial elements – walls, floors, places for gathering, places for sitting – but the ways in which we use those elements will change. Light, invisible, networked technologies can permeate familiar public spaces without modifying their physical appearance. A great shift in the scope of urban design will be a focus on transforming the cities we already have, rather than designing new spaces or structures. Digital transfusions might make a system more efficient (such as optimized public transit), more environmentally sustainable (new modes of food production) or simply more sociable (offering new networks and ways of physically meeting people). The basic premise is "more silicon, not more asphalt."

6. The harnessing of data and its increasing visibility and accessibility is a powerful tool to generate awareness regarding urban issues. How can this be used to resolve some of these issues? How can data promote behavioral changes to foster civic engagement?

C.R. + M.C.: Big data is a way of seeing the true, real-time patterns and modes of habitation of people living in a city – it can reveal what we call "the signature of humanity." But in order for big data to have a big impact, it needs to be analyzed and digested. Today, that is done by visualization experts, who might make an image, a graph, or a video; what we are now working on at the lab is making a tool that puts data at your fingertips. Massive amounts of information will be readily accessible and intuitively navigable. When everyone – from scientists to politicians to everyday commuters – has the opportunity to explore invisible dimensions of the city around them, we will see radical behavioral change. When the city is revealed, particularly in real time, we all will become smart citizens.

1. http://senseable.mit.edu/trashtrack/ (Accessed: 15 June 2018).

2. http://www.hubcab.org/#13.00/40.7250/-73.9484 (Accessed: 15 June 2018).

TRASH | TRACK

Elaborated by the SENSEable City Lab and inspired by the NYC Green Initiative, Trash I Track focuses on how pervasive technologies can expose the challenges of waste management and sustainability. Trash I Track uses hundreds of small, smart, location aware tags: a first step towards the deployment of smart-dust-networks of tiny locatable and addressable micro-eletromechanical systems. These tags are attached to different types of trash so that these items can be followed through the city's waste management system, revealing the final journey of our everyday objects in a series of real time visualizations. The project is an initial investigation into understanding the 'removal-chain' in urban areas, and it represents a type of change that is taking place in cities: a bottom-up approach to managing resources and promoting behavioral change through pervasive technologies.[1]

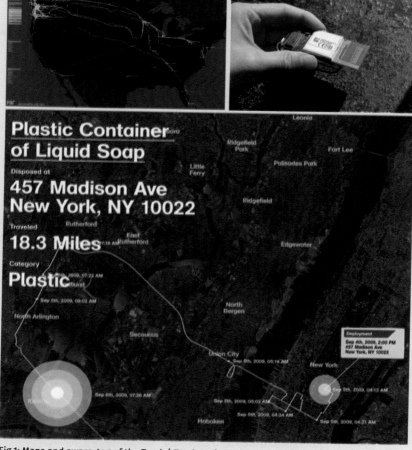

Fig.1: Maps and aware-tag of the Trash | Track project.

HUBCAB

With an ever-increasing trove of real-time urban data streams, we are able to see precisely where, how, and at what times different parts of our cities become stitched together as hubs of mobility. By using these pervasive, interconnected, and smart technologies, we can begin to unravel the complexity of our travel patterns and identify how we can reduce the social and environmental costs embedded in our transportation systems. In HubCab we target taxicab services as a way to understand the linkages between our travel habits and the places we travel to and from most often. HubCab is an interactive visualization that invites you to explore the ways in which over 170 million taxi trips connect the City of New York in a given year. This interface provides a unique insight into the inner workings of the city from the previously invisible perspective of the taxi system with a never before seen granularity. HubCab allows to investigate exactly how and when taxis pick up or drop off individuals and to identify zones of condensed pickup and dropoff activities. It allows you to navigate to the places where your taxi trips start and end and to discover how many other people in your area follow the same travel patterns. What do these visualizations tell us about collective mobility? How many of these cabs might you have been able to share with the people around you? And how might entertaining these questions be the first step in building a more efficient and cheaper taxi service?[2]

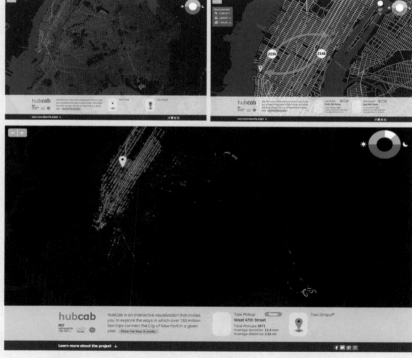

Fig.2: Density maps of collective mobility.

03

Systemic Architecture

ecoLogicStudio

1. How do you explain the concept of a systemic approach to architectural design, and how is this embedded in your research projects?

e.S.: The notion of systemic design begins by engaging a different set of relationships within urban space. At the beginning of our research, we defined a city as an extended field of interactions that cannot be separated from a larger context. Rather than understanding cities as isolated entities, we have recently started to look at cities as nodes that are part of extended networks of energy, information, and objects. This rethinking of spaces and systems in our research and practice provides us with an understanding of dynamic entities that can be developed through mechanisms that influence the city's morphology by operating across regimes and scales.

In this regard, we started to look at everything in the city, especially its infrastructure, spatial, a-biotic, micro-scale systems to understand constant relationships within the urban environment, the inhabitants, and other systems. In this systemic analysis we gain feedback we can use to determine ways to interact, intervene, or manipulate the city. This is probably the essence of why we are interested in a systemic approach to architecture that we define as "systemic design practice." This new, more democratic production process, defined by analysis and feedback loops, is replacing the old hierarchy of expertise. The efficiency of design solution is built up through revisions. Intelligence and performance are incorporated as a consequence.[1]

2. Your book *Systemic Architecture* provides a toolbox of design protocols organizing your experiments in three categories: environments, machines, and behavioral systems. How does each of these categories propose a particular research and design approach to systemic architecture?

e.S.: The three categories were developed from the attempt to systematize a diverse range of projects that have been developed in the last few years considering different material systems, scales, and contexts. The ideas of environmental machines and behavioral systems are part of a methodological diagram built by these three categories. We understand the city as extended territory of operational fields:

First, the representation of diagrammatic pre-urban structures is developed to allow cross-fertilization of different types of informal fields and urban stimuli within a coherent urban plan. Second, such coherent spatiality of multiple stimuli is a precondition for the developments of protocols for the occupation of new territories or the redevelopment of existing urban landscapes; within this operational framework, ecologic feedback, participation and social self-organization may happen. Third, operational fields require a diagrammatic type of representation; associative and algorithmic modeling techniques are at the core of this diagrammatic urban machine.[2] These machines focus on the material, morphological, spatial, and performative dimensions of the interlacement of systems. This research in behavioral systems informs both teaching and practice.

As the city will be embedded by relational mechanisms, its conception evolves into a kind of simulation. However the simulated virtual city is not a model of a generic digital urban landscape or a reductive abstraction of the real city; rather, it is a meta-space in which the contingent and the accidental are absorbed, and the generic is developed through the organization of differences and variations. Then the actualization of the self-organizing city manifests itself as a progressive acquisition of material and performative specificities in relation to a specific contextualization process. A diverse pool of actors, agents, forces, and protocols cooperate to form and reform the urban space. As such, urban space defined as behavioral is the product of processes of co-evolution of multiple agents behaving as a coherent assemblage.[3]

3. Some of your projects connect different ecological and social systems in what you call "proto-architectures." How do your projects attempt to engage the notion of the city as a territory of self-organization?

e.S.: Proto-architecture emerges from going beyond the typological understanding of architecture. A systemic thinking applied to the city requires a redefinition of the idea of typology that must include machinic systems. This approach does not refer to the "mechanical machines" but rather tries to embrace what we call the "machinic paradigm." In that sense we are looking at proto-architecture as prototypical assemblages of different devices that are technological, ecological and aesthetic.

When you try to identify those systems of co-evolution, you will need to identify a platform to establish a new ground that operates within the notion of emergence. You have different sources of information in the city which enable those strategies to evolve and a form of self-organization to operate. The self-organizing city is a new form of an emergent real-time world city, a conceptual and operational model of urbanization that promotes the re-structuring of engendered species of social, economic, and environmental practices and organizations. By harvesting this inherent vitality, the self-organizing city proposes a vision of urbanity that has interaction and a narrative of productive know-how derived from its constitutional protocols; as such, the self-organizing city has no limits either by time or by space, no beginning and no end, no fixed and final configurations, no permanent dweller, no single planner. Within the self-organizing city, these forces generate diversity and cultivate differentiation as means to evolve true novelty. Inhabiting the self-organizing city gives us the ability to play an active role in the making of an open urban future, turning destruction and erasure into potential for new originations, transformations, and migrations.[4]

4. In your projects, the notion of prototype and protocol become intrinsically connected and turn into operational tools to design and plan contemporary cities. How can both foster new models to address our global ecological crisis?

e.S.: As Gilles Clement says, "If we look at the earth as a territory devoted to life it would appear as an enclosed space, delimited by the boundaries of living systems (the biosphere). In other words it would appear as a garden. This statement impels every human being, in its transient existence on earth to commit to social responsibilities as guarantor of the living world. Humans have received the world to manage it and to become a gardener."[5]

We often refer to the writings of Gilles Clement for his interesting definition of the difference between the landscape architect and the gardeners. According to Clement, the gardener always has his hands dirty and continuously deals with the process of life. He does not achieve that through conjectures, but he does it out on the field. This is also the way we would like to imagine the processes that shape the urban design and architectural practice. The approach connects directly to the notion of protocol. A key idea of systemic understanding of architectural design is to reject the boundries of our milieu that we have set for ourselves.

The protocol is thinking about the design practice as an operational practice that is constantly engaged with a particular type of environment and context. Machinic protocols embed technology into material organization that become part of everyday ecological practices of planetary cultivation of the biosphere, operating at multiple interconnected scales.[6]

In this scenario, prototypes become important vehicles for testing these relationships. In our projects, prototypes are defined as contingent assemblages of a large number of components organized via multiple relationships. The emergent properties of these assemblages exceed the sum of their constituent parts. Regulated and evolved through feedback loops of interactions, prototypes differentiate in the lineages that progressively develop specialist and dedicated behaviors, form and actual material organization.[7]

5. The notion of Machine has evolved, and it is now changing its relationship with technology and systems. You borrow the term machinic from Deleuze and Guattari.[8] What is the potential of machines or machinic processes as powerful tools to envision future scenarios?

e.S.: One of the key aspects is to overcome the mechanical paradigm and to understand machinic processes as learning systems. Self-learning systems are already widely implemented in robotics, artificial intelligence, and gaming, but clearly architectural processes are still struggling with it. In our work, we try to apply the concept of biological machines to a spatial, urban, and infrastructural level. We are often very sceptical about "green" projects because some of them still operate within a mechanical paradigm and perform within infrastructural systems that are hierarchical and controlled. In our research practice, we are developing strategies to generate bio-energy using only living systems and catalytic processes investigating forms of transforming centralized systems into networks and distributed intelligence. Infrastructural networks are more rhizomatic then hierarchical. This aspect is directly linked to social and political aspects. We have to redesign those infrastructures with that in mind if we really want to move forward. The idea of systemic and machinic will then engage reality.

Our design diagrams can be embodied into material and architectural prototypes. Under such a condition, architecture performs as an analogue computer, reading contextual forces, loading programmatic and material constraints, and feeding back spatial, ornamental, and behavioral solutions. By definition such an architectural mechanism is deeply ecological as it is defined only in relation to a specific habitat, or a specific ecology of material, forces, information, flows, and energetic fields. We call that an ecoMachine to underline how different such a conception of the term "machine" is when we move away from the mechanical paradigm. Such a philosophical shift has a profound material impact on our understanding of the architectural assemblage and on the role of technology within it. It forces a dissolution of the boundaries between the inert (structural frames and envelope) and active or responsive (machinic systems), between users and their surrounding environment, between the body and the architectural space, and between the furniture and the city.[9]

6. The Fish&Chips Apparatus project brings up an interesting concept for generating different forms based on real-time data. How do you expect this to be fitted into future projects as well as into the broader field of architecture?

e.S.: The Fish&Chips Apparatus was an experiment to develop a design interface. The challenge was to integrate within a project both the soft real-time and the actual real-time. Architecture is typically less concerned with time than space, but from the new systemic perspective, time becomes an essential ingredient of any algorithmic design protocol. In our projects, we have identified two distinct time frameworks: soft- and real-time. We define the process of digital simulation and manufacturing of a project as soft-time processes and the live running of a project as real-time processes. Both are conceptually separated. For the Fish&Chips Apparatus project, we were working with a lagoon in Dubai, rethinking an architecture that could be connected to the marine life.

7. How can your study of systems potentially inform future architectural and urban scenarios? How do you envision your tactics evolving into larger, long-term strategies? How do you imagine these prototypes to become a part of architectural components?

e.S.: This is ongoing research, and some of these questions are part of our continuous experimentation. To be able to implement these experiments at a larger urban scale, there are many aspects that need to change first. This includes political and governmental control. Some projects are pushing these boundaries through means of research. The Micro-Algae farming project, for instance, is not only about algae but about the understanding of a biological organism as part of what Gilles Clement calls "Third Landscape," an ecosystem that is not completely codified. A further question is, how can we attribute a code to this urban ecology and learn from it? We are interested in the development of prototypical spaces to make us aware of new possibilities. Eventually some of them will evolve into a much larger scale project. Others will remain more temporary.

STEMCLOUD V2.0

The STEMcloud v2.0 project involves the public in an oxygen-producing machine. The machine is a breeding ground for micro-ecologies existing in the Guadalquivir, a river of Seville, Spain. The public feeds the colonies present in the river water with nutrients, light, and CO2. As a result oxygen is produced in the gallery space which triggers the growth process though its interaction with the public and in turn affects the growth patterns with its visual effects. The transparency and porosity of the architectural system exposes the process to be visually and materially interfering with the microclimate of the gallery. The gallery environment and the river are used as components of a system to provoke different feedback circles within the city of Seville.

This extended model of systemic architecture can be understood in cybernetic terms as a multilayer crossing of changing and transforming feedback loops that define the qualities of a new ecologic understanding of architecture. As architects, we only define the starting point or the primed condition of the project. The project itself emerges from the interaction with the public.[10]

Fig.1: Detail and overall views of STEMcloud v2.0.

H.O.R.T.U.S.

The project seeks the notions of urban renewable energy and agriculture through a new gardening prototype. This proto-garden hosts micro- and macro-algal organisms as well as bioluminescent bacteria and is fitted with ambient light-sensing technologies and a customized virtual interface. The project stimulates the emergence of novel material practices and related spatial narratives. Flows of energy (light radiation), matter (biomass, CO_2) and information (images, tweets, stats) are triggered during the four-week-long growing period, inducing multiple mechanisms of self-regulation and evolving novel forms of self-organization. H.O.R.T.U.S. proposes an experimental hands-on engagement with these notions, illustrating their potential applicability to the master planning of large regional landscapes and the retro-fitting of industrial and rural architectural types. The biologic diversity within the project is provided by lakes and ponds within Central London. As algal organisms require CO_2 to grow, visitors are invited to contribute by blowing air inside the various containers (photo-bioreactors) as well as adjusting their nutrients' content. Oxygen is released as a result, feeding the other organisms in the briccole (bioluminescent bacteria) and in the space. Information flowing daily through H.O.R.T.U.S. feeds its emergent virtual garden and is accessible via smart phones. Its virtual plots are nurtured by the flow of observations posted by each visitor, locally and globally, by lighting levels, data streams, and by human interaction in real time.[11]

Fig.2: View of H.O.R.T.U.S. and mobile interface.

META-FOLLIES

META-follies for the Metropolitan Landscape is a spatial mechanism that establishes a playful dialogue with its users. The project creates clouds of knowledge as a real-time meta-conversation that is based on material experience, pattern recognition, and sedimentation of feelings.

The project has been conceived algorithmically and will keep evolving algorithmically once handed over to its end users. It will commence its exhibition tour following unknown itineraries across different regions, cultures, languages, and environmental contexts. We cannot predict how these contexts will read and respond to the project nor if the pavilion can be designed to respond to a specific context in a traditional sense. However the machinic nature of this proposal allows for a dynamic contextual relationship with the surrounding urban environment through intuitive aesthetic appreciation and behavioral response. Referring to Slavoj Zizek's call for the development of a "new terrifying form of abstract materialism," the pavilion confronts the artificiality of the contemporary urban landscape with the production of a new form of hyper- artificiality able to offer refuge and consolation to the crowd of post-ecologists.[12]

Fig.3: Views of META-follies pavilion.

FISH&CHIPS APPARATUS

The Fish&Chips Apparatus project is part of a larger concept about cyber gardening which asserts that machines, humans, environment, and computers will co-design and co-evolve into new artificial living systems. Fish produce small and high frequency waves as they swim. Sensors record the wave frequncies in real time. Parametric modeling software reads the wave patterns and generates an ever-evolving design output. The digital gardens drawings turn into interactive maps of fish colonies' daily behavior, of the visitors' feeding patterns, and the designer's reaction to both of them.[13]

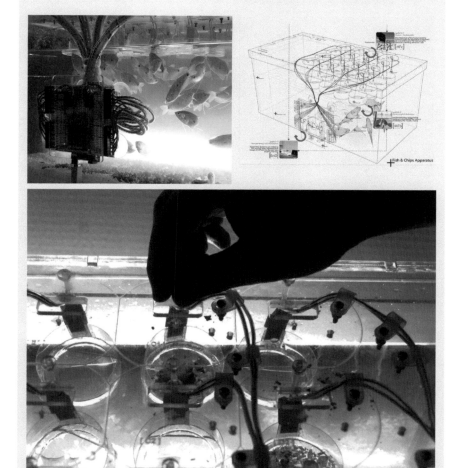

Fig.4: View of the Fish&Chips apparatus.

AQVA GARDEN

Aqva Garden, is an artificial garden that functions as a distributed rain collector and as a water storage system. Unlike conventional recycling systems, the project doesn't hide its functional apparatus; rather, it embodies it in its structural matrix of a branching system. Moreover it operates by expanding the climatic effects latent within the management of water and its transitional states such as evaporation. Rain water becomes the protagonist of perceptual games and gardening processes, opening new potentials in the conception of ecologic infrastructures for the built environment.[14]

Fig.5: Aqva Garden, a rainwater collection system.

1. Pasquero, Claudia; Poletto, Marco. *Architecture as Systemic Practice* in Systemic Architecture. Routledge. 2012, pg. 11.

2. Pasquero, Claudia; Poletto, Marco. *Operational Fields* in Systemic Architecture. Routledge. 2012, pg. 27.

3. Pasquero, Claudia; Poletto, Marco. *Behavioural spaces* in Systemic Architecture. Routledge. 2012, pg. 27.

4. Pasquero, Claudia; Poletto, Marco. *The Ecology of the Self-Organizing City* in Systemic Architecture. Routledge. 2012, pg. 3.

5. Clements, G., Il Giardiniere Planetario, 22Publishing, Milan, 2008, pp.58-59

6. Pasquero, Claudia; Poletto, Marco. *Coding as Gardening* in Systemic Architecture. Routledge. 2012, pg.7.

7. Pasquero, Claudia; Poletto, Marco. *Machines* in Systemic Architecture. Routledge. 2012, pg.115.

8. Kampis George, *Self-Modifying Systems in Biology and Cognitive Science. A New Framework for Dynamics, Information and Complexity*, Pergamon Press, Oxford, England, 1991, p. 235.

9. Pasquero, Claudia; Poletto, Marco. *Relational ecological Machines* in Systemic Architecture. Routledge. 2012, pg.185.

10. http://www.ecologicstudio.com/v2/project.php?idcat=7&idsubcat=17&idproj=35 (Accessed: 15 June 2018).

11. http://www.ecologicstudio.com/v2/project.php?idcat=7&idsubcat=71&idproj=115 (Accessed: 15 June 2018).

12. http://www.ecologicstudio.com/v2/project.php?idcat=3&idsubcat=66&idproj=120 (Accessed: 15 June 2018).

13. http://www.ecologicstudio.com/v2/project.php?idcat=7&idsubcat=17&idproj=55 (Accessed: 15 June 2018).

14. http://www.ecologicstudio.com/v2/project.php?idcat=3&idsubcat=17&idproj=15 (Accessed: 15 June 2018).

04

Cities, Data, Participation and Open Source Architecture

Usman Haque

1. Engaging the regular citizen in the collective production of urban space seems to be a main concern in your work. Pachube, an on-line database service provider developed by your firm, allows citizens to connect sensor data to the Web and to build their own applications. The project takes advantage of open data to discover and share information. How will the increasing access to real-time data change the design process and as a consequence the future of public space?

U.H.: There are two important aspects of capturing real-time data that interest me. The first one: I don't think that meaning is found in the pure data, but I do believe that meaning is created through the way data is measured. You get a wider range of different meaning depending who, how, why, and where the data was measured. There is not necessarily any consensus on how to measure, but what emerges is a mode of analyzing and perceiving the world through multiple perspectives.

The second aspect relates more directly to architecture. In conventional models the design phase is distinct from the construction phase and distinct from the phase of a building being occupied. Usually the architect, the engineer, or the designer is working with a set of parameters, what I call a "presumed data set." When you move on to the construction phase, you almost discard the presumed data set because you are in the design implementation phase. By the time the process gets to the occupancy phase, all the initial assumptions dissipated throughout the process makes it impossible to compare the result to the initial assumptions.

What really intrigues me in engaging real-time data in the design process is the possibility of introducing a feedback loop throughout the entire process: controlling the performance of the construction phase in real time, and more importantly during the occupancy when you can actually run occupancy evaluations challenging the assumptions you made at the beginning. This would close the feedback loop, returning to the initial "presumed data set." To summarize: the first aspect that I am interested in is this notion of multiplicity provided by the open data; the second one is the capacity for the designer to implement the feedback loop beyond the design process.

2. In the pamphlet "Urban Versioning System" you talk about the possibility of an open source urban design. How could open source code enable people to become more consciously involved in the design, production, and inhabitation of urban space?

U.H.: In this pamphlet that I wrote with Matthew Fuller, we apply the idea of open source to the design processes of urban space. The idea behind open source systems allows everyone to change, reconfigure, and adapt a design.

In open source processes, there are two aspects to consider. One is the potential of open sourcing the source code (genotype); two is open sourcing the artifact itself (phenotype). This increases participation and responsibility and the empowerment of citizens. Open source will then as a consequence affect the management and configuration of urban space.

The pamphlet provides strategies for open source urban design. One of them is titled "Build Rather than Design." Here design and planning are abandoned in favor of starting construction immediately. All design activities can be performed on the actual material while the thought process directly engages the artifact.[1] This principle emerges from the idea of taking ownership of both the spatial configuration itself and the future occupancy and use at the same time.

This led to another idea that we presented in the pamphlet: designers often need to abandon the preciousness connected to their design in favor of the potential of seeing their work taken apart and reassembled through different logics. In software applications such a preciousness does not exist; instead, almost a sense of pride emerge when new possibilities arise from this process of disassembling and reconfiguring.

3. In your work, you talk about the concept of "granularity" as the resolution for varied levels of participation to a system. How does that work? What is the potential of this concept in relation to the possibilities of the construction of an urban space designed by citizens?

U.H.: As a system designer, I am interested in the involvement of the citizen in the construction and creation of the urban experience. An example is the design of a simple light switch. The idea of *"granularity"* there would suggest to engage the user in questioning the simple on-off switch and design and program new ways of controlling lighting. I often apply this idea of granularity at multiple levels with the goal of creating multiple entry points and allowing for interaction at many different levels.

4. You describe architecture as an operating system, in particular referring to the relationship between "soft-space" (programs that animate the machine) and "hard-space" (the physical machine). What is the relationship between the two, and how have you investigated that through your projects?

U.H.: Hard spaces are the immutable aspects of architecture such as floor, ceiling, or walls. Soft spaces are more mutable aspects such as sound, light, smell, temperature, or social relationships. If we consider architecture as the operating system uniting the two, a strong potential emerges from the exploration of the relationships between the creators of the system and the users.

In the mid-90s I started applying the concept of operating systems to architecture. I was intrigued about artifacts that could be created by the same people using them, situations in which producers and consumers were the same. To conceptualize architecture that way allows me to define new frameworks for responsibility and ownership where boundaries between the definition of hard and soft space are increasingly blurred.

5. If we consider the shift of the notion of connectivity from the object to the environment, how do your public interventions engage an urban context and participatory processes?

U.H.: One of the projects that engages those relationships is Open Burble, which was developed for Singapore Biennale in 2006. The main concept was to actively engage the members of the public to construct public space itself. Often our conception of urban space is that somebody else produced it, and we just live in it. Open Burble is a system that allows the public to design and build everything by itself with a support of directed inputs. The project was built in one day. After completion, it was 15 stories tall, creating a strong impact on the urban skyline.

The people were active participants, able to control the system by changing the position of handles at the bottom of the structure. They could control its movement, color distributions, and flows. Its changing appearance was a result of bottom-up processes of individual citizens negotiating with each other and with environmental conditions such as the direction and speed of the wind. The process was able to empower people through a sense of agency and ownership while reversing traditional top-down processes.

PACHUBE

Pachube is a Web service available at pachube.com that enables you to store, share, and discover real-time sensory and environmental data from objects, devices, and buildings around the world. It is a platform that contributes to build the Internet of Things. Its key aim is to facilitate interaction between remote environments, both physical and virtual. Apart from enabling direct connections between any two responsive environments, it can also be used to facilitate many-to-many connections, just like a physical "patch bay" (or telephone switchboard). It enables any participating project to plug into any other participating project in real time so that, for example, buildings, interactive installations, or blogs can "talk" and "respond" to each other.[2]

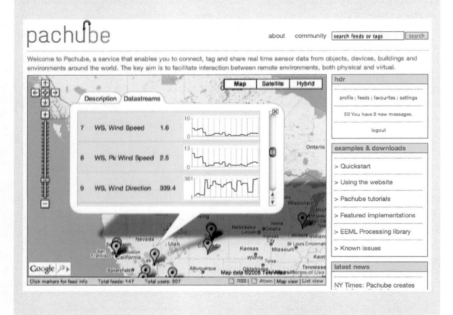

Fig.1: Wep page of Pachube.

DIY City 0.01a

DIY City 0.01 is a prototype for a future mass participation performance developed in collaboration with Special Moves. In the project, streets and building facades become the canvas for three-dimensional projections at an urban scale. The spatial wiki allows citizens to reimagine and redesign urban space.[3]

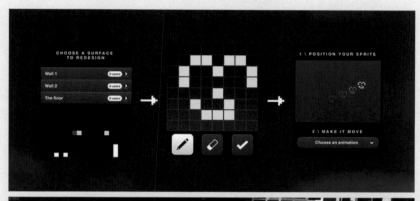

Fig.2: Processing devices and interface of DIY City 0.01.

MARLING

In Marling citizens become players on an urban stage, bringing to life a large out-door public space through their actions and sounds and building shared public memory of collaboration. Citizens generate three-dimensional effects using their voices, forming delicate, intricate, and colorful animated spatial enclosures within the crowd. With the use of lasers, microphones, and custom-built software, Marling enables thousands to witness and participate in the collective experience of the construction of an "aurora" at an urban scale.[4]

Fig.3: Animated public environment of Marling installation.

NATURAL FUSE

Natural Fuse, creates a city-wide network of electronically assisted plants that act both as energy providers and as circuit breakers to prevent carbon footprint overload. A power socket enables people to charge their electronic devices while the plant's growth offsets the carbon footprint of the energy expended. The amount of energy expended in the system is balanced by the amount of CO_2 absorbed by the plants growing in the system. By networking them together, the units can "borrow" excess capacity from other units not currently being used to share their capacity and take advantage of carbon-sinking-surplus in the system as a whole, since it is likely that not all fuses will be in use at any one time. If people cooperate on energy expenditure, the plants thrive (and everyone may use more energy while remaining carbon neutral). If they don't cooperate, the plants start to die, thus diminishing the network's electricity capacity as a whole.[5]

Fig.4: Network of electronically assisted plants.

OPEN BURBLE

Open Burble is constructed from 140 configurable carbon-fiber modules approximately 2 meters in diameter. Each module is supported by seven extra-large helium balloons (for a total of about 1000 individual pixels) containing sensors, LEDs, and microcontrollers that enable the modules to coordinate amongst each other to create patterns of color that ripple up towards the sky. The people on the ground manipulating the Burble's handles see their actions echoed as colors and movements through the entire system. They see their individual fragments, perhaps even identifying design choices they have made. Their individual contributions become an integral part of the group experience. Part installation, part performance, Open Burble enables people to contribute at an urban scale to a structure that occupies their city.[6]

Fig.5: Open Burble against Singapore skyline.

1. Haque,U. Fuller. Urban Versioning System 1.0. The Architectural League of New York, New York. 2008.

2. Haque Design + Research. Pachube. http://www.haque.co.uk/pachube.php (Accessed: June 15, 2018)

3. Haque Design + Research. DIY City 0.01a. http://www.haque.co.uk/diycity.php (Accessed: June 15, 2018)

4. Umbrellium. Marling. http://umbrellium. co.uk/portfolio/marling/ (Accessed: June 15, 2018)

5. Umbrellium. Natural Fuse. http://umbrellium.co.uk/portfolio/natural-fuse/ (Accessed: June 15, 2018)

6. Haque Design + Research. Open Burble. http://www.haque.co.uk/openburble.php (Accessed: June 15, 2018)

05

Situated Technologies

Omar Khan

1. In your essay "Interaction Anxieties," featured in the book _Sentient City: Ubiquitous Computing, Architecture, and the Future of Urban Space_,[1] you talk about the need to "expand our understanding of what interactivity could be as computation becomes more pervasive requiring a shift away from the instruments of interactions—and towards the relations we expect to achieve from them. These include the ways in which we communicate and socialize with one another and inhabit our cities and the world." How could this be enabled in the future? How does the shift "towards a more phenomenological framing of information as situated, contextual and embodied" lead to a more sentient environment?

O.K.: My major interest in that article was to recast the problems of interaction in terms that were provocative for architects and designers of interactive and responsive environments. This is different from the way technology companies see interaction, which has to do with making communication between humans and computers "seamless." In my opinion, technologists take too literally Mark Weiser's statement that "the most profound technologies are those that disappear. They weave themselves into the fabric of everyday life until they are indistinguishable from it."[2] It is not their literal disappearance that is of significance but how they "weave" themselves into social life. Ubiquitous computing focusses entirely on the former: how to make the technology disappear into the ether and allow for mediated communication to become as easy as human conversation. But as we know, the introduction of any technology into everyday life, whether visible or invisible, fundamentally changes the ecology of social engagement. As McLuhan says, there is always an amputation with any technological extension.[3] So what concerns me is not the physical invisibility of these technologies, which is in fact wonderful, but that their weaving into everyday life is made invisible from public engagement and scrutiny.

This provides an opportunity for architects and designers to shift the conversation about information technologies away from augmentation and seamless access to where they must make a difference: meaningful social engagements. This will require more open methods for culling the vast territory of Big Data, and my argument is this cannot be done solely through technological means. Information will need to be situated, become more contingent on context, and develop a body that allows people to meaningfully engage, navigate, and use it. This can't simply happen through algorithms but will require interfacing those algorithms with physical materials, environmental controls, and the physical networks of the built environment. This is what I see as the phenomenological framing of information where architects, designers, and artists are and can continue to make a serious contribution.

In addition, a truly sentient environment can only result through greater mutualism between humans, technologies, and ecologies. This requires that all are present, as Bruno Latour contends, through a Parliament of Things in the public sphere.[4] This is the potential problem posed by current trends in pervasive computing which see in technological disappearance an argument for better communication rather than a lack of representation. However, I think the adoption of these technologies should be more contentious because then we are confronted with making decisions more publicly and democratically. It allows us to address concerns with surveillance, access, and control that we may be willing to concede to technologies if we are given the choice.

2. What role do situated technologies play in bridging the gap between the pervasive network and the physical built environment?

O.K.: They are small gatekeepers in the flow, like Maxwell's demons resisting entropy, but also meaningfully tying the local to the global. Situated technologies provide a means for an individual or a small group to act within the vast information environment. Their use needn't be limited to people. Objects and buildings can also be given agency to contribute through them. One opportunity this provides is representation in the public sphere. By contributing local information, they can provide higher resolution of data tied to specific contexts. This enriches the variety of sources and quality of information. This also provides access to local actors and empowers them to act with greater confidence in their context.

The significance for architecture is that embracing these technologies provides an alternative way, besides the orchestration and catalyzing of spectacle, to contribute to the built environment. Architecture has done a very good job of providing spectacle, like the Bilbao effect, to affect change in the built environment. But this can only work in exceptional cases and is heavily tied to large capital. Responsive architecture and situated technologies provide new avenues for buildings, public spaces, and inhabitants to participate in the public sphere. As these technologies move closer to becoming pervasive like utilities, we really should be exploring new architectural programs that can take advantage of them.

3. How will the insertion of embedded, mobile, pervasive technologies in the public space, affect governmental policies (i.e. privacy laws, zoning, etc.) and urban development?

O.K.: The smart city is already present, and it has a clear program for improving policies and guiding urban development. Its concerns are tied to minimizing redundancy, conserving resources, and better managing risk so that capital can keep flowing. This is a solution that many cities yearn for, including the one I reside in, Buffalo, NY. Unfortunately, the better management promise of the smart city comes at a price which is homogeneity, where "best practices" will be uncritically imposed on every constituency to conform to standards that the technologies are capable of addressing. My colleagues and I at the Center for Architecture and Situated Technologies are very critical of the "smart city" but also see the opportunities it can provide for the emergence of the "sentient city."[5] In the latter, the same technologies are conscripted to open the management of the smart city to democratic participation, empowering bottom-up actions and increasing the variety of responses to pressing problems rather than homogenizing responses.

4. What is the potential of information technology as a catalyst for social, political, and economic change?

O.K.: It fundamentally has to do with access. Pervasive and accessible information technologies can be the basis for leveling the field of participation in human affairs, the public sphere, and the economy. We might think this is really the role of business and government legislation, but I hold little regard for either when it comes to fostering the public sphere. Public space is one means of access to the public sphere, and architects, designers, and artists can continue to work on new conceptions for the accessible public space.

5. How would situated technologies transform civic public life?

O.K.: The biggest transformation recently of civic life by a technology has come from the introduction of the mobile phone. Some effects have been less than ideal like an increased focus to the device at the expense of experiencing the context, exchanging glances with passers-by, and being fully in the public place. But there are also opportunities this technology provides to connect with friends and likeminded people while navigating the city in new ways. What of course would be ideal is if greater serendipity, chance, and provocation could be introduced into these meetings and movements. This is where situatedness of a technology is very important. A project we did with media artist Osman Khan called SEEN: Fruits of your Labor for the Zero1 Festival in San Jose suggests a possible alternative. It introduced a situated technology—a curious urban screen—to a public plaza. The screen is made from infrared LEDs, hence invisible to the naked eye but visible through cameras on people's mobile phones. It projects responses that can take the form of texts, images, and videos from the public to the question, "What is the Fruit of your Labor?" The screen became an attraction but also a forum for opening a public dialogue. Part of this had to do with facilitating individual opinions to become public, and part of it was to make the receiving of the messages fun and not didactic. Many unexpected public performances resulted, including sharing phones with strangers, inter-generational discussion explaining what was happening, and spontaneous crowd formations. These draw people out from the personal space of their devices and into productive public exchanges.

6. How are situated technologies embedded in your practice becoming a tool for design?

O.K.: Situated technologies are primarily a concept/attitude and secondarily actual technologies (sensing, actuating, computing, and fabricating). The latter are probably part of most architects' toolkits, with perhaps sensing and actuating technologies still new to some. In my practice, they have provided tools to more clearly address issues of response, duration, and performance in design. Projects like Open Columns and Gravity Screens employ a variety of material and electronic technologies to develop responsive structures. Their intention is to choreograph the occupancy of space. Open Columns ties carbon dioxide sensing with a retractable columnar structure that deploys to change the space of occupancy. Gravity Screen affects space by changing the movement of people within it from linear to polar. In both cases, it is the effect on the behavior of occupancy that concerns me most. But to alter behavior, you have to have the occupant complicit in it. It can't be something done to them, otherwise it quickly becomes predictable and no better than a Disney ride. The hardest part in developing such technologies is that lack of precedence and good practices. But that is probably what makes it so engaging.

OPEN COLUMNS

Open Columns is a system of nonstructural columns that resides collapsed in the ceiling of a space. They are made from composite urethane elastomers and can be deployed in a variety of patterns to reconfigure the space beneath them. These patterns create gradations of enclosure, either in plan through the full deployment of columns, in section through their partial unfurling to change ceiling heights, or through a combination of the two. The system is a mutable architecture that can change the perception and inhabitation of the space within which it is deployed. At its most trivial, the columns can be preprogrammed to deploy themselves in prescribed configurations. This can be effective for re-proportioning a large space into smaller spaces or reorganizing the circulation of people through it. A more complex program ties the columns to real-time sensing such that they can respond to inhabitants' perturbations in space. The columns, working from a simple set of rules, respond to data coming from a carbon dioxide (CO_2) sensor. In a reasonably enclosed environment, CO_2 values can radically change with the inclusion of people. The columns are programmed to come down when CO_2 levels are going up, resulting in people dispersing into smaller groups. If the CO_2 levels are going down, the columns respond by going up, effectively inviting people into the space. If however the CO_2 value stays static, the columns cycle through a random set of configurations until the CO_2 either goes up or down. The relation between the changes of the CO_2 level and the configuration of the columns provides information that is stored and used as feedback by the system. In this way the columns, over time, learn about their space based on their own actions within it. This creates a teleonomic environment, one that acts on particular goals but has no determinate goal to which it is ultimately driven.[6]

Fig.1: Open Column project and model.

GRAVITY SCREEN

Gravity Screen is a surface construction whose morphology results from gravity's effect on its material patterning. It is composed from two elastomers of different Shore hardness that take an organized form when the screen is hung. Rubber's elasticity and high weight to volume ratio make it particularly problematic as a self-supporting material. However, the compounded effect of excessive weight on a stretchable material results in it stiffening. By crisscrossing hard and soft rubbers, Gravity Screen uses this property to create a controlled stretch. The hard rubber acts as a cross brace to the soft rubber, creating a changing surface weave that has structural properties. Traditionally, screens use modularity to maintain pattern continuity so that when they are repeated, the module literally reflects the collective. Gravity Screen's modules have more nuances since their individual behaviors affect not only the look of the entire screen but its structural and formal properties. The screen's half arch design exhibits one formal variation of this type of building system. Divided into columns, each module has the same depth (numbers of rubber layers) with different barcode pours. These pours iteratively go from thin to wide along the structure's surface to create the semi-arch form. The resulting tight and loose hexagonal patterns on the module faces negotiate material tensions and ease into their final shape with the assistance of gravity.[7]

Fig.2: View of Gravity Screen.

1. Shepard, Mark (ed.).*Sentient City: Ubiquitous Computing, Architecture and the Future of Urban Space.* MIT Press. 2011.

2. Weiser, Mark. *The Computer for the 21st Century.* Scientific American. September 1991.

3. McLuhan, Marshall. *The Medium is the Message.* Understanding Media. 1964.

4. Latour, Bruno. We have never been modern. Harvard University Press. 1993.

5. Shepard, Mark (ed.). *Sentient City: Ubiquitous Computing, Architecture and the Future of Urban Space.* MIT Press. 2011.

6. http://cast.b-ap.net/opencolumns/form/ (Accessed: June 15, 2018)

7. http://cast.b-ap.net/reflexivearchitecturemachines/gravity-screen/ (Accessed: June 15, 2018)

06

Synthetic Urban Scenarios

Areti Markopoulou - Manuel Gausa

1. As stated in "The Metapolis Dictionary of Advanced Architecture: City, Technology and Society in the Information Age," 21st century forms of architecture emerge by layered conditions that engage technologies, networks, data and ecologies as well as socio, political and economic forces. How do you envision the role of the architect to change in the near future? What emerging technologies will drive that change?

A.M.: Today we are facing a paradigm shift in the field of architecture and design due to the impact of information and communication technologies. Architecture as a practice has always been in close connection with technology but in the current era technology is evolving at a very high speed, and this is increasingly transforming the way in which we produce spaces. As we are challenged by global urbanization and economic crisis, these conditions make us question methodologies and techniques to facilitate these types of processes.

Architects have the responsibility to open up to new paths that rely more on collaboration and new forms of networked practices that operate on multi-disciplinary platforms. The architect must be able to deal with multiple scales simultaneously and propose visions on how we inhabit spaces, cities and the whole planet. The capacity to understand and use existing tools is not enough; on the contrary, we need to start creating our own tools based on emerging needs.

For instance, if we talk about manufacturing processes in architecture and design, the capacity to produce new tools brings together many different disciplines. The emergence of open source and collaborative design challenges the architect to learn and work more with the users, different disciplines and forms of practice. The feedback loop between learning and making is a very important component to shift boundaries to new modes of working. In this framework, prototyping represents an important mode for experimentation that exists between learning and creating your own tools for design.

M.G.: The current paradigm shift emerges primarily from new ways in which we can process information. The new model overrides the deterministic modernist and the calligraphic postmodernist model to propose an emergent way to register, perceive and act on the urban space that is complex, layered, transversal and ultimately information-based.

The big revolution is the enormous capacity of information to interact and exchange with inter-scalar systems. The capacity of fostering different forms of interaction has always been present, but now we are experiencing a condition of simultaneity of information that triggers conditions of larger complexity and a new way to act on the reality. This information-based interaction is the big revolution of our time. A product of information-based interaction is the emergence of relational spaces where conditions of exchange are much more complex and move beyond deterministic and taxonomic logics.

2. What are the current approaches of implementing information and communication technologies in different layers of the urban environment? How can this trigger new categories of projects, technologies and solutions that could be extended systematically to cities in different geographic regions thus helping them to become more efficient and more human?

A.M.: First of all I would like not to use the word 'smart' cities but instead propose a model that is more sentient, responsive and adaptable. The emergent urban model is composed both by visible and invisible conditions; for instance, data is an invisible layer of information that both users and environments produce. This is influencing in real-time the way in which the city is functioning. To be able to capture these dynamic conditions we have to start designing in a new way.

Responsiveness is key to understand these dynamic models. We are witnessing a change in which the capacity to sense, track and monitor urban data is shifting the way we design, for instance, infrastructures. Now we have the capacity to design systems that are able to self-program themselves and monitor/respond to real-time data. Those data become generative. Locative media and smart devices are also radically changing the way people receive information and interact with it.

We continuously receive, collect and exchange data with the urban environment through continuous feedback loops. The fact of being able to somehow control and understand how information is transmitted and used will help us to make more informative and responsible decisions on how we inhabit urban space and interact with each other.

M.G.: In the present conditions we understand that the relations between cities and forms have changed. We question approaches that are more formal and/or typological in favor of more connected, systemic and layered systems. This multi-layered condition opens up to the operative notion of simultaneity. The more layers in the system, the more exchange, and the more dynamic conditions in the urban space can be registered. This dynamicity has increased its speed.

The city is a complex and evolutionary system. As architects we work constantly with these conditions that are inherently non-linear and heterogeneous. In this framework, new technologies that are the catalysts of this change constantly embrace uncertainty as a productive design category. The more exchanges we have in a system, the more paradoxes are generated.

These relations are not only about the human and the machine but more importantly with the continuous exchange with systems around us such as the environment, infrastructures, resources and so on. The typological breeding is the type of operation that is occurring; this generates new categories of projects and forms that test possible futures for the built environment.

3. How can we move towards more sustainable, open and user-driven ecosystem to boost future internet-enabled services of public interest and citizen participation?

A.M.: I believe that this question relates strongly with the new cities model. Information and communications technologies, apart from creating new categories of projects, are also creating new categories of city-based services. For instance, in 2014 Barcelona was awarded the European Capital of Innovation; the award reflected on how the city was able to re-structure and propose a vision for service-oriented infrastructures for the citizens.

Barcelona has also been one of the first pioneers in promoting fablabs as public services located in different city neighborhoods but connected in a service-based network. This represents the rise of a new form of soft-infra-structure that emerges locally but impacts the city as a whole. Those systems decentralize services and are both localized and part of the expanded network. There are many models similar to this that are impacting the way in which we access services in cities, from co-working platforms and crowed-sourced services to other platforms such as Uber and Airbnb.

M.G.: As architects we not only design forms, but more importantly, we try to foster human habitats through the design of relations in space. In this regard, public and collective space is not only a monumental/iconic/representational/ functional space but it's an active space where technologically-informed devices can foster the rise of a new type of relational space expressed through collective forms.

This means that we are constantly witnessing the emergence of a new type of active public space and sensible relational environments. These conditions of collectiveness open up to a different type society where the city becomes more of a city for citizens. New complex inter-scalar spatial devices and mechanism have the capacity to compress and synthetize those conditions in active rela-tional spaces that are simultaneously locally situated and globally networked.

4. How do you foresee the emergence of a new urban economy and management models created through efficient, responsive, decentralized, and hyper connected systems be implemented in order to build the city of the future?

A.M.: There are two ways in which those implementations are occurring: one more top-down and one more bottom-up. The two are not exclusive but instead complementary. The ultimate goal is not the top-down application of technologically mediated systems but on the contrary the rise of a more human-driven urban environment.

We have to be aware both of the human scale as well as policies that need to be provided to lead us to the creation of strategies for how things could be implemented.

The definitions of policies and collaborative decision making processes are crucial to make a real impact on the city. At the same time everything should start from understanding the needs and users instead of the pure top-down application of what we call 'smart.'

M.G.: To frame the question one important thing to consider is the re-localization and new distribution of governance nodes within the city. The capacity to create a network of physical nodes that are able to process data, information and so on challenge us to negotiate the more localized, partially individualized nodes with tensions produced by the larger, holistic, global network.

How governance and policies are developed and deployed is a very important aspect of these emergent forms of urban management and economies. In this context, architects and urban designers can contribute more by giving strategic/programmatic visions for cities and their complex set of relations.

5. In the current design practice, as educators, how can we teach to envision projects that integrate public spaces, buildings, infrastructures, user interactions and information technologies?

A.M.: I believe this starts from how we can create those connections, educate each other and be active participants in current issues we are facing in the field of architecture and design. We should focus more on how we can design relations and less on how we can design forms. Forms eventually will emerge as outcome of those relations.

We have to keep in mind that not only technology can generate new models, but can also re-integrate with analog forms of creating new relations to produce urbanity. The analog and digital ultimately are always interdependent.

M.G.: We are in a moment where the architect is more an agent able to synthetize ideas to project scenarios. We have to propel explorations in our field through continuous research and testing. Architecture schools could be and should be platforms for experimentation rather than only the consolidation of what we consider disciplinary knowledge. I believe this is not only about the transmission of professional skills but the awareness of the evolving conditions of what it means to be an architect in our current expanded practice.

Vythos

Over 70 percent of the planet's surface is covered by water, the majority of which is in the world's seas and oceans. About 99 percent of all freshwater ice resides in two ice sheets: Antarctica and Greenland. Both are expected to melt if humanity's CO2 output isn't curbed quickly. Since 1880, global sea levels have already risen by 8 inches (200 mm). Sea levels could rise an additional 1.3 meters (4.3 feet) in the next 80 years. Not only are sea levels rising; the rate of their rise is increasing. Half of the world's population lives within 60 km of the sea, and three-quarters of all large cities are located on the coast. Global flood damage for large coastal cities could cost $1 trillion a year if cities don't take steps to adapt. Barcelona is one of these cities. Vythos is an ambitious project that seeks to raise awareness for this issue by including the citizen's direct contribution in a real-time experience. It not only spreads awareness on how climate change affects our environment, it allows us to measure citizens' knowledge about this important subject and encourages them to offer possible solutions. The prototype is located along the Espigon del Gas, one of the central piers of the famous beach of Barceloneta. The hope is that Vythos will be replicated in other important coastal cities that will be impacted by rising sea levels.

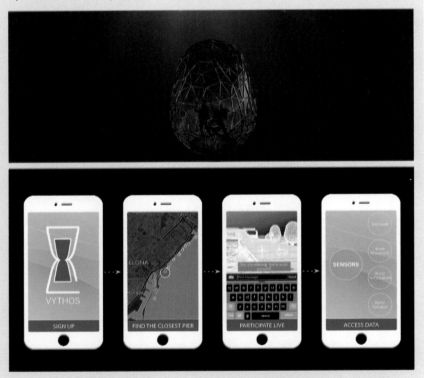

Fig.1: View of the Vythos prototype and App. (Project by Alex Mademochoritis and Laura Marcovich, Master in City and Technology, Open Source Urbanism Seminar, Faculty: Marcella Del Signore, Fall 2016, IaaC, Barcelona).

Canvas

The Canvas project proposes to question an existing component of a facade, such as louvers, in the city of Barcelona in order to turn this element into support of digital art creations. The aim is to make a digital art museum at the scale of the whole city. The physical Canvas surface hosts a QR code which refers to the digital art production the user has uploaded on the global interconnected database. Each QR code is unique and refers to a specific digital content with a specific geo-localization. The Canvas app can be downloaded by anyone and allows users to see the production of digital artists at the specific location of their Canvas code. Once the app is able to read the code, it provides to the user an augmented reality scene within the city streets where the digital content of the creator is integrated to the urban context. This new digital augmented reality layer applied onto the actual physical city is accessible through the camera of your connected devices using the Canvas app. Canvas finds an application for anyone who wants to share digital art content as part of a community.

Fig.2: View of the Canvas app on a facade in Barcelona (Project by Iacopo Neri and Sylvain Totaro Master in City and Technology, Open Source Urbanism Seminar, Faculty: Marcella Del Signore, Fall 2016, IaaC, Barcelona).

07

Machines and Apparatuses as Scenario

François Roche

1. In your projects, you use machines and apparatuses. How do they inform your projects? What is the story behind your Alchimis(t/r/ick) machines?

F.R.: At the Alchimis(t/r/ick)[1] college there are some machines- some desirable machines- that love to pretend to do more than they really should do. In a pursuit of pataphysics, the science of imaginary solutions, they never reveal their inner nature, their origins and illusions, genuineness and fakeness. Simultaneously speculative, fictional, and accurately and efficiently productive, they navigate in the world of *Yestertomorrowday*, happily and innocently, walking briskly over the mountain of 20thcentury rubbish. Using strange apparatuses, these Alchimis(t/r/ick) machines symmetrically articulate different arrows of time and layers of knowledge, but more specifically they negotiate the endless limit of their own absurdity, where behavior that seems illogical is protocolized by an extreme logic of emerging design and geometry, where input and output are described by rules and protocols.

Neither a satire of the worlds, a techno-pessimism, nor a techno-derision, they are located at the limit--or constitute the limit--between the territory of conventions, certainties and stabilities where one can comfortably consider everything legitimated by an order, or an intuition of an order, and all other territories, whether produced by paranoia or fantasy or reported back by travelers.

2. Is there a difference between your machines and other conventional machines, and if there is, what makes your machines unconventional?

F.R.: In a casual and basic sense, machines have always been associated with technicism and used as the extension of the hand, through its replacement or improvement by accelerating its speed and power to produce and transform. But it seems very naïve to reduce machines to this first, obvious layer of their objective dimensions, in a purely functional and "machinism" approach, exclusively limited to Cartesian productive power, located in the visible spectrum of appearance and facts. Machines simultaneously produce artifacts, assemblages, multiplicity, and desires and infiltrate the *raison d'être* of our own body and mind in the relationship to our own biotopes.[2] Basically everywhere in nature, they are at the origin of all processes of exchange, transactions of substances, entropy, and vitalism.[3] Machines are a paradigm for the body in the sense of its co-extensibility with nature, through processes, protocols, and apparatuses, where transitory and transactional substances[4] constitute and affect simultaneously all species, their identities, their "objectivized and subjectivized" productions, and their mutual relationships.

3. What is the main difference between mechanic operations and what you define as "machinic" operations?

F.R.: In this pursuit of a polyphonic approach, we cannot overlook the concept of the "bachelor machine" [5] as an attempt to integrate "machinism" apparatuses into a narrative of transaction and transmutation (in the alchemical sense). Contradictorily, these 'Alchimis(t/r/ick)-machines operate as direct critique and denunciation of capitalist managerial reductionism, which replaced uniqueness and rarity with a system of repetition and standardization, erasing both the workers (when they are not becoming machines themselves[6]) and any singularities, any anomalies, providing products for a strategy of servitude which combined mass production and the production of the alienation of mass, as described by Walter Benjamin.[7] In opposition to this predictable one-way dependency, bachelor machines simultaneously convey the fascination of this sophisticated human construction, its barbarian eroticism,[8] the impulsion and repulsion it generates as a permanent schizophrenia alternating between its simultaneous potential for production and for destruction,[9] for a permanent dispute between Eros and Tanatos. They are vectors of both, resistance and production, infiltrating the arrogance of the mainstream and revealing its schizoid values. The same industrial system produces both outcomes; their geneses are consubstantial, and their diametrically opposed collateral effects depend mainly on our ability to see and make visible what lies beyond the mirror.

4. Your practice has been developed as a continuous exchange between fiction, speculation, and research. Within this framework, how will those parameters be generative of architectural realities?

F.R.: In the work of R&Sie(n), 'Alchimis(t/r/ick) machines try to reveal these disturbances, or are constitutive of them. The blurriness between what they are supposed to do--as perfect alienated and domesticated creatures--and the anthropomorphic psychology we intentionally project on them, creates a spectrum of potentiality, both interpretative and productive, which is able to re-"scenarize" the operating processes. A mind machine simultaneously transforms the real and our perception of what we consider real. In this sense machines seem to be vectors of narratives, generators of rumors, and at the same time directly operational, with an accurate productive efficiency. These multiple disorders, this kind of schizophrenia, could be considered a tool for reopening processes and subjectivities, to "re-protocol" indeterminacy and uncertainties. Agents of blur logic, of reactive and reprogrammable logic, the scenario created by and through these "machinism" processes asymptotically touch their own limits, revealing the fragile and movable borderline between what seems to be, what should be, and what should have been. The creatures produced by this machinism confront exterior forces, their ambivalence, their contingency, their instability. They allow us to "exercise our power, to be conscious of our power, that is by the same token self-consciousness, consciousness of our vulnerability in the face of the enormity of this power."[10]

5. Are your machines a critique on our society and our relationship to our natural and built environment?

F.R.: Genetically Siamese and consubstantial, this stuttering appears as a dysfunctional reflection in the mirror, organizing the way "we and the others" frame conflicts arising precisely from the state of the mirror, to quote Lacan: "Where the perception of the unicity of our corporality, through the mirror, is constructed in coincidence with the defragmentation of the perception of our environment."[11] The process of "reductionism" to One Body is the symmetrical reflection of the One World, where all the complexities, the schizoid and paranoid assemblages, early childhood's sweet disruption of consistency, are trapped in a univocal representation, framed and simplified. And consequently all the alien fragments that cannot fit in this perfect and comfortable representation of "INselves and OUTselves" are considered fatally flawed by absurdity, weirdness, and oddity to preserve the illusion of this symmetrically operative but vain unicity.

'Alchimis(t/r/ick) machines seek to articulate things and minds, objective and narrative production, "machinism" causalities and unpredictable dependences, to interrogate their *raison d'être* and the eroticism of their transgression, weaving together the *malentendus* and the illusions they generate, in a different arrow of time: "Here and now" as a live transaction, "here and tomorrow" as an operative fictional scenario, "elsewhere and simultaneously" for speculative and political research, navigating between apparatuses of "animism, vitalism, and mechanism."

The tools of mechanization drift from a self-organized urbanism (*an "architecture des humeurs"*)[12] to a stochastic machine with a predictable completion (*Olzweg*),[13] from the "machinism" ghost of a wild DMZ forest (*heshotmedown*)[14] to a paranoiac uranium laboratory (*TbWnD*),[15] to a simple transportation machine, a stargate experiment (*Broomwitch*).[16]

OLZWEG

Olzweg is a stochastic machine that vitrifies the city, starting from a museum of architecture as the origin of the transformation and contamination. A smearing of recycled glass is operated through a process of staggering, scattering, and stacking. This random aggregation is a part of an unpredictable transformation, as in Franz Kafka's "Metamorphosis." The starting point is known, the outcome unpredited, as a fuzzy logic of the vanishing point. The machine extends a museum acquiring "voluntary prisoners" wrapped in a permanent entropy of the graft. The opposite of an architecture that petrifies, historicizes, panopticalizes, classifies, and freeze-dries in the maze of its multiple trajectories.

Fig.1: Machine processing and interface of Olzweg.

HE SHOT ME DOWN

The he-shot-me-down machine is a tracked biomass machine that penetrates into the demilitarized zone between North and South Korea, collects the rotten substances, the superficial coating of the forest in decomposition, and brings back this material to plug all the external surfaces of the building. In this way it is fermenting grass to create a natural eco-insulation. The machine collects the pathological ingredients of the period of demilitarisation and recycles them for productive use. This no-man's land has been abandoned since the end of the war more than half a century ago. In this re-appropriation of nature by nature, elves, wizards, witches, and harpies come back, new species appear, and legends and fairy tales are transported back to the safe zone, the south zone, as in a "Stalker" experiment.

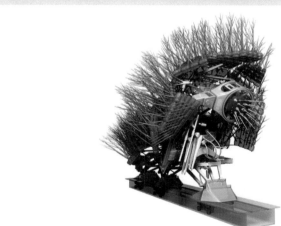

Fig.2: Views of he-shot-me-down.

TBWND

TbWnD is an alert machine or a marker of our past and future. It is a laboratory of dark adaptation and detection of solar radiation intensity. It detects the afterglow on external surfaces, the influence of the sun's seasonal and daily emissions on surfaces directly "touched" by its rays. The phosphorescent components operate as a UV sensors. The intensity of UV rays hitting the area and its occupants during daytime is indicated through lighting effects during nighttime. Its glass components reveal the danger of the sun's radiation and the decreasing ozone in the stratosphere. This machine articulates the danger of men's impact to nature and becomes a marker for the uncertain future of human development's after-effects.

Fig.3: Views of TBWND.

BROOMWITCH

Broomwitch is a machine for the Seroussi family house in Meudon, France. It is an extension of an André Bloc house, a domestic museum filled with paintings, a cabinet of curiosities. The machine is a parasite in the garden of the house, creating stimuli of against nature. Its hydroponic tentacles are colonising "heterotopically" the back of the garden. Two sun glass eyes are perforating the two turbulence vortexes formed by its tentacles.

Domesticating and quietly gradually violating nature, the machine allows for artistic substitution by transitioning from green turbulence to smooth white walls to support its production of art.

Fig.4: Diagrams and views of Broomwitch.

AN ARCHITECTURE DES HUMEURS

A utopian machine to produce a self-organized urbanism conditioned by a bottom-up system that uses multitudes (in Spinoza's and Antonio Negri's use of the word) to drive the entropy of its own assembly and construction system. As *architecture des humeurs*, the machine is based on the potential of a recently developed technology to reread human corporeality through its physiology and chemical balance. The machine uses this technology to make the emotional transactions of the "animal body," the headless body, and the body's chemistry palpable and perceptible, informing us about individuals' adaptation, their sympathy and empathy, and their reactions at being confronted with a particular situation and environment. The machine translates this information through its "machinism" behavior as an endless construction process of secretion and weaving. This process can generate a vertical structure using 3D printing for extrusion and sintering and bio-plastic cement as hybrid raw material that chemically agglomerates to physically constitute the computational trajectories. This structural calligraphy works like a machinism stereotomy comprised of successive geometries according to a strategy of permanently repeting anomalies.

Fig.5: Installation process of Architecture Des Humeurs.

HYBRID MUSCLE

Hybrid muscle is a project conceptualized as a story by Philippe Parreno and François Roche in Chiang Mai, Thailand. The story titled "The Game" is about the creation of Siamese twins. In the game "Hybrid Muscle," a shelter produces a movie, and "Boys from Mars," a movie, produces a shelter. The movie by Philippe Parreno and the shelter by R&Sie(n) are contingent and simultaneous, and tenants and artifacts of the game.

Fig.6: Exterior and interior view of Hybrid Muscle.

1. A reference to the *Collège de 'pataphysique* founded in 1948 in honor of Alfred Jarry. We could consider the OuLiPo (*Ouvroir de littérature potentielle*, Workshop of Potential Literature, whose members included the mathematician François Le Lionnais and Raymond Queneau) the first branch of that group, and the 'Alchimis(trick)' as a rotten branch of that branch.

2. A reference to the work of Ilya Prigogine, who considered human beings a "mechanism" of exchanges, of shared substances IN and OUT and vice versa.

3. "Vitalism presumes a monadological rather than atomistic ontology. In Leibniz's 'monadaology' all substances are different from one another, whereas its opposite, Cartesian 'atomism,' presumes that matter is comprised of identical parts (atoms)." Scott Lash.

4. "All bodily phenomena can be explained mechanically or by the corpuscular philosophy." Leibniz, *Letters to Arnauld*.

5. Developed by a multitudes of artists, philosophers and writers such as Duchamp, Poe, Kafka, Deleuze and even, subconsciously, Cervantes. The term "bachelor machine" was first used by Marcel Duchamp around 1913 in connection with pieces of work that would later be assembled in the Large Glass of 1915-1923. For Deleuze and Guattari, the "bachelor machine" forms a knot between desiring machine and the body without organs, to create a new myth which seems to articulate Narcissus, Opheus and Sisiphus. It has been isolated by Michel Carrouges (in his book *les Machines célibataires* / Arcanes, 1954)

6. As seen in the organization of Henry Ford's Detroit factories (Fordism).

7. Walter Benjamin, "The Work of Art in the Age of Mechanical Reproduction," 1936.

8. From Poe's "The Pit and the Pendulum" and Kafka's "In the Penal Colony" to Ballard's *Crash*.

9. The latest subculture icon: *Avatar*, where the metempsychosis machine saves the ecologically-balanced Ewya Kingdom from the caterpillar machine which destroys the blue hobbits' dreamtimes. The both are coming from the same "tea pot".

10. Recorded lecture by Gilles Deleuze at the University of Paris-Vincennes in 1980. The exercise of our power as Nietzsche and Deleuze understood it, as a gift, a creation, and not the kind of dominance that they (the machines) could grant us.

11. *Le stade du miroir. Théorie d'un moment structurant et génétique de la constitution de la réalité, conçu en relation avec l'expérience et la doctrine psychanalytique* 1937.

12. *An "Architecture des Humeurs"* / 2010

13. *Olzweg* / 2006.

14. *Heshotmedown* / 2009.

15. *The Building which never dies* / 2009 -10.

16. *Broomwitch* / 2008.

ACKNOWLEDGEMENTS

We would like to thank those who have contributed directly to the book by providing information and images about projects and case studies or participated in the conversation section of the book with interviews.

We would also like to express our gratitude to the institutions where we currently teach, the New York Institute of Technology School of Architecture & Design and the New Jersey Institute of Technology College of Architecture and Design, for their support. We also thank other academic institutions in the USA and Europe where we taught during the early research stages for this book.

In addition, we would like to thank Christine Rhine for her support in the copy-editing phase of the book. Finally, we would like to thank our past and present colleagues for the encouragement and inspiring conversations during the different phases of the development of the book and to thank family and friends for their continuous support.

ILLUSTRATION CREDITS

Introduction:

- Fig.1: Active Public Space – Glories Regenerative Systems; Image courtesy of Edouard Cabay and Rodrigo Aguirre. Active Public Space – Glories Regenerative Systems is a project of IaaC - Institute for Advanced Architecture of Catalonia, developed in the Master in Advanced Architecture in 2015/2016 by: Students: Peter Geelmuyden Magnus, Utsav Mathur, Tobias Deeg, Martin Hristov, Rana Abdulmajeed, Nour Mezher, Jean Sebastian Munera, Lili Tayefi; Faculty: Edouard Cabay, Rodrigo Aguirre with help from: Angel Muñoz, Pablo Barquin, Carmen Aquilar Wedge, Ramin Shambayati.

- Fig.2: Archigram, Walking City,1964; Cedric Price, The Fun Palace,1964; Archigram, Instant City, 1964. Commissioner: Joan Littlewood, Socially Interactive Learning and Recreational Space, Lea Valley, London 1961 – 1966, published in World Architecture News, http://www.worldarchitecturenews.com/project- images/2012/21461/cedric-price/reader-review-fun-palace.html?img=1

- Fig.3: Urban Acupuncture, 2007; Image courtesy of Jaime Lerner. Atelier d'Architecture Autogérée, Tactics for Resilient Post-urban Development, 2014; Image courtesy of Atelier d'Architecture Autogérée.

- Fig.4: Kinetic Wall, Prototype exhibited at the Venice Biennale 2014; Image courtesy of Barkow Leibinger.
- Fig.5: Augmented Structures v1.1.: Acoustic Formations; Image courtesy of Alper Derinboğaz and Refik Anadol.
- Fig.6: Under scan, 2005; Image courtesy of Rafael Lozano-Hemmer.
- Fig.7: People's Canopy, 2015; Image courtesy of People's Architecture Office.

Essays:
Interacting
- Fig.1: Defensible Dress, 2001; Image courtesy of Höweler + Yoon Architecture.
- Fig.2: Skinput, 2010; Image courtesy of Chris Harrison, Carnegie Mellon University.
- Fig.3: The Brasserie Restaurant, 2000; Image courtesy of Michael Moran.
- Fig.4: The Cultural Shed, 2011; Image courtesy of Diller Scofidio + Renfro.
- Fig.5: Prada Transformer, 2009; Image courtesy of OMA-Rem Koolhaas.
- Fig.6: Gardens by the Bay, 2011; Image courtesy of Grant Associates.

Integrating
- Fig.1: Integrated Spatial layers; Image courtesy of G. Riether and M. Del Signore.
- Fig.2: Tube Exists App by Wavana.
- Fig.3: Integrated Urban Systems; Image courtesy of G. Riether and M. Del Signore.
- Fig.4: Stadse Boeren App, www.stadseboeren.nl
- Fig.5: Pokémons, www.pokemon.com
- Fig.6: Mindmixer App, www.mindmixer.com

Expanding
- Fig.1: Bosch mobility solutions and automated driving; Image courtesy of Bosch.
- Fig.2: Studies for Hyper-Reality, 2013; Image courtesy of Keiichi Matsuda.
- Fig.3: Tidy Street, funded by the EPSRC; Image courtesy of Yvonne Rogers.
- Fig.4: Urban Augmentation App; Image courtesy of Yuichiro Takeuchi and Ken Perlin.
- Fig.5: Responsive robotic arm; Image courtesy of Christian Moeller.
- Fig.6: Urban Tapestries: Contexts Map, courtesy of Urban Tapestries: Contexts Map (image: Proboscis 2006).

Hacking

- Fig.1: Polizei lügt; Image courtesy of Peter Weibel.
- Fig.2: Oasis Nr. 7, Haus-Rucker-Co / Laurids Ortner, Günter Zamp Kelp, Manfred Ortner and Klaus Pinter, documenta 5 / Kassel 1972.
- Fig.3: Environment Transformers, (Flyhead, Viewatomizer, Drizzler) Haus-Rucker-Co, Laurids-Zamp-Pinter, Vienna 1968. Photo: Gerald Zugmann 1968.
- Fig.4: Mind Expander, Haus-Rucker-Co / Laurids-Zamp-Pinter / Vienna 1968.
- Fig.5: Hacking Traffic Lights by Ztohoven; Image courtesy of Roman Týc.
- Fig.6: Building Yourself an Urban Reserve,1998; Image courtesy of Santiago Cirugeda.
- Fig.7: Air Quality Egg; Image courtesy of Air Quality Egg.
- Fig.8: Smart Citizen do-it-yourself Kit; Image courtesy of IaaC- Institute for Advanced Architecture of Catalonia.

Networking

- Fig.1: Vision of global electrical network by Buckminster Fuller.
- Fig.2: Network diagrams based on a 2015 analysis by Mc Kinsey Global Institute.
- Fig.3: Map of Velib, based on information from Velib App.
- Fig.4: Audi Urban Future; Image courtesy of BIG – Bjarke Ingels Group.
- Fig.5,6: Amphibious Architecture, 2009; Image courtesy of The Living, David Benjamin.
- Fig.7: WikiPlaza, by José Pérez de Lama, Sergio Moreno Páez, Laura H. Andrade; Image courtesy of Hackitectura.

Case Studies

- Eco-boulevard - Air Tree by Ecosistema Urbano; Image courtesy of Ecosistema Urbano.
- Cosmo by Andrés Jaque / Office for Political Innovation; Image courtesy of Imagen Subliminal, Miguel de Guzmán.
- Datagrove by Future Cities Lab; Image courtesy of Future Cities Lab, Select Photography: Peter Prato.
- Wendy by HWKN; Image courtesy of HWKN.
- iLounge by Marcella Del Signore (X-Topia) + Mona El Khafif (ScaleShift); Image courtesy of Marcella Del Signore and Mona El Khafif.
- Lumen by Jenny Sabin Studio; Image courtesy of Jenny Sabin Studio, Photography: Yuriy Chernets.

- Living Light by The living; Image courtesy of The Living, David Benjamin.
- iWEB by ONL/ Kas Oosterhuis; Image courtesy of Kas Oosterhuis.
- Bubbles by FoxLin; Image courtesy of Michael Fox.
- Skin by Gernot Riether + Damien Valero; Image courtesy of Gernot Riether and Damien Valero.
- Cloud by Single Speed Design; Image courtesy of Single Speed Design.
- Blur by Diller Scofidio + Renfro; Image courtesy of Diller Scofidio + Renfro.
- Serpentine Gallery Pavilion by OMA; Image courtesy of OMA.
- Digital Water pavilion by Carlo Ratti Associati; Image courtesy of Carlo Ratti Associati.

Conversations
Urban Social Design, Network, Participation and Open Source City
Ecosistema Urbano
- Fig.1: View of bioclimatic improvement strategies for public spaces; Image courtesy of Ecosistema Urbano.
- Fig.2: Air Tree Shanghai; Image courtesy of Ecosistema Urbano.
- Fig.3: Energy Carousel; Image courtesy of Ecosistema Urbano.
- Fig.4: Interior of Galicia Pavilion; Image courtesy of Ecosistema Urbano.

SENSEable CITY
Carlo Ratti with Matthew Claudel
- Fig.1: Maps and aware-tag of the Trash | Track project; Image courtesy of Carlo Ratti.
- Fig.2: Density maps of collective mobility; Image courtesy of Carlo Ratti.

Systemic Architecture
ecoLogicStudio
- Fig.1: Detail and overall views of STEMcloud v2.0; Image courtesy of ecoLogicStudio.
- Fig.2: View of H.O.R.T.U.S. and mobile interface; Image courtesy of ecoLogicStudio.
- Fig.3: Views of META-follies pavilion; Image courtesy of ecoLogicStudio.
- Fig.4: View of Fish&Chips apparatus; Image courtesy of ecoLogicStudio.
- Fig.5: Aqva Garden, a rainwater collection system; Image courtesy of ecoLogicStudio.

Cities, Data, Participation and Open Source Architecture
Usman Haque
- Fig.1: Wep page of Pachube; Image courtesy of Usman Haque.
- Fig.2: Processing devices and interface of DIY City 0.01; Image courtesy of Usman Haque.
- Fig.3: Animated public environment of Marling installation; Image courtesy of Usman Haque.
- Fig.4: Network of electronically assisted plants; Image courtesy of Usman Haque.
- Fig.5: Open Burble against Singapore skyline; Image courtesy of Usman Haque.

Situated Technologies
Omar Kahn
- Fig.1: Open Column project and model; Image courtesy of Omar Kahn.
- Fig.2: View of Gravity Screen; Image courtesy of Omar Kahn.

Syntetic Urban Scenarios
Areti Markopoulou - Manuel Gausa
- Fig.1: View of Vythos prototype and App; Image courtesy of Alex Mademochoritis and Laura Marcovich.
- Fig. 2: View of the 'Canvas App' and implementation on Barcelona's facade; Image courtesy of Iacopo Neri and Sylvain Totaro.

Machines and Apparatuses as Scenario
François Roche
- Fig.1: Machine processing and interface of Olzweg; Image courtesy of François Roche.
- Fig.2: Views of he-shot-me-down; Image courtesy of François Roche.
- Fig.3: Views of TBWND; Image courtesy of François Roche.
- Fig.4: Diagrams and views of Broomwitch; Image courtesy of François Roche.
- Fig.5: Installation process of "Architecture Des Humeurs"; Image courtesy of François Roche.
- Fig.6: Exterior and interior view of Hybrid Muscle; Image courtesy of François Roche.

MARCELLA DEL SIGNORE

Marcella Del Signore is an associate professor at the New York Institute of Technology School of Architecture and Design. She is the principal of X-Topia, a design-research practice that explores the intersection of architecture and urbanism with digital practices. She holds a Master in Architecture from University La Sapienza in Rome and a Master of Science in Advanced Architectural Design from Columbia University in New York.

Her work concentrates on the relationship between architecture and urbanism by leveraging emerging technologies to imagine scenarios for future environments and cities. The notion of urban-digital prototyping has been at the core of her research. She currently serves on the board of directors of ACADIA (Association for Computer Aided Design in Architecture) where she is technical co-chair for the 2018 conference in Mexico City.

She has taught and collaborated with academic institutions in Europe and the US including Tulane University, Barnard College at Columbia University, the Architectural Association, Institute of Advanced Architecture of Catalonia, University of Waterloo, LSU School of Architecture, IN/ARCH- National Italian Institute of Architecture and University of Trento. At Tulane University, she served as the director of the study abroad program in Rome and as the Kylene and Brad Beers SE Professor at the Taylor Center for Social Innovation and Design Thinking. She has received several awards and has lectured, published and exhibited widely.

GERNOT RIETHER

Gernot Riether is the director of the School of Architecture and associate professor at the College of Architecture and Design at the New Jersey Institute of Technology. An international lecturer, he previously taught at Kennesaw State University, Ball State University, ENSA Paris La Villette, Georgia Tech, the New York Institute of Technology and Barnard College at Columbia University. He holds a DI in Architecture from the University of Innsbruck in Austria, and a Master of Science in Advanced Architectural Design from Columbia University in New York.

Riether's research explores the relationship between public urban spaces and information technology. His research has been funded by governmental institutions, nonprofit organizations, the construction industry and universities. Projects he designed and built with his students in his Digital Design Build Studio have won competitions and are featured in many books on digital fabrication. Riether is the author of over 40 refereed papers, articles and book chapters.

He serves on the board of directors of the Consortium for Sustainable Urbanization, a non-profit organization affiliated with UN-Habitat and UN ECOSOC, and on the board of directors of ACADIA (Association for Computer Aided Design in Architecture) and is editor of the *Journal of the Design Communication Association*.

URBAN MACHINES
Public Space in a Digital Culture

Author
Marcella Del Signore
Gernot Riether

Editorial Director LISt Lab
Alessandro Martinelli

Published by
LISt Lab
info@listlab.eu
listlab.eu

Art Director & Production
Blacklist Creative, BCN
blacklist-creative.com

ISBN 9788898774289

**Printed and bound
in the European Union,**
October 2018

Collection BABƎL

Sales, Marketing & Distribution
distribution@listlab.eu
listlab.eu/en/distribuzione/

LISt Lab is an editorial workshop, based
in Europe, that works on contemporary
issues. LISt Lab not only publishes, but
also researches, proposes, promotes,
produces, creates networks.

LISt Lab is a green company committed
to respect the environment. Paper, ink,
glues and all processings come from
short supply chains and aim at limiting
pollution. The print run of books and
magazines is based on consumption
patterns, thus preventing waste of
paper and surpluses. LISt Lab aims at
the responsibility of the authors and
markets, towards the knowledge of a new
publishing culture based on resource
management.